P9-CCS-711

HARVARD HISTORICAL STUDIES • 144

Published under the auspices
of the Department of History
from the income of the

Paul Revere Frothingham Bequest
Robert Louis Stroock Fund
Henry Warren Torrey Fund

David Paull Nickles

Under the Wire

How the Telegraph Changed Diplomacy

Harvard University Press
Cambridge, Massachusetts, & London, England 2003

Printed in the United States of America

Library of Congress Cataloging-in-Publication Data

Nickles, David Paull, 1966–
Under the wire : how the telegraph changed diplomacy / David Paull Nickles.
p. cm. — (Harvard historical studies ; v. 144)
Includes bibliographical references and index.
ISBN 0-674-01035-3 (alk. paper)
1. Telegraph–History. 2. Diplomacy–History. 3. Diplomats–History.
4. Negotiation–History. I. Title. II. Series.

HE7631.N516 2003
327.2′09′034—dc22 2003051113

Contents

Introduction

It is axiomatic in the world of diplomacy that methodology and tactics assume an importance by no means inferior to concept and strategy.

—*George F. Kennan*

Electric telegraphy—based on the idea of converting messages into charged impulses that could be sent over wires at lightning speed to distant receiving stations—was the first great step into the world of telecommunications. In 1830 there was no telegraph industry. By 1910, combined telegraph traffic in just four countries (Great Britain, the United States, France, and Germany) was more than a quarter of a billion telegrams annually.[1] What this number meant in human terms is impossible to express: the wires carried news of deaths, births, triumphs, and disasters. In the short run, the telegraph had an enormous effect upon journalism, financial markets, and the popular imagination. In the longer run, it greatly influenced military strategy, writing style, and business organization. The telegraph industry propelled many workers from modest backgrounds into middle-class status, and it created important job opportunities for women. At the same time, many (though far from all) telegraphers toiled in highly regimented conditions. The early telecommunication industry created skilled, technologically savvy, geographically mobile, but sometimes restive groups of laborers who often felt alienated from their employers. In these respects, telegraphers were much like the computer programmers and "information technology" workers of today. And like their more recent counterparts, telegraphers suffered from repetitive stress injuries. "Carpal tunnel syndrome" was once called "telegrapher's paralysis."[2]

Telegraphy produced significant challenges and opportunities for those involved in the conduct of diplomacy.[3] After all, international relations, at their core, involve the conveyance of data across international boundaries. Telegraphy, a technology that recast long-distance communication, had the potential to transform diplomacy. How did governments respond to the new possibilities it offered? The question gains added drama if one considers that diplomats tended to favor the conservation of past practices. This is the central question underlying this book: What happened when skeptical, somewhat recalcitrant consumers—whether conservative ambassadors or custom-bound bureaucrats—encountered a potentially revolutionary medium of communication?

Three Themes

In order to illuminate the influence of telegraphy upon international diplomacy, I focus on three themes: the changing degree of autonomy available to diplomats; the effects of speed on diplomacy and the lives of diplomats; and the effects of the technical and economic characteristics of telegraphy on foreign policy establishments. Each issue forms the basis for one of the book's three parts.

Part I begins with an examination of the origins of the War of 1812, a subject that illuminates the role of diplomats during the era before electric telegraphy. This story introduces a question: Did telegraphy give governments greater control over distant diplomats? Many contemporaries believed so. Some argued that telegraphy centralized authority to such an extent that it necessitated drastic reform in diplomatic institutions. Telegraphy, they insisted, eliminated the need to spend money on permanent diplomatic representatives in foreign countries. This contention implied that the telegraph had transformed diplomats into automatons. Diplomats added nothing to the conduct of international politics, and the public would not suffer if statesmen or foreign ministries dispensed with intermediaries and communicated directly with one another. There is some truth to this argument, and one can find evidence in its support. At the same time, the effect of telegraphy was more complex than this argument implies. First of all, one should not assume that the oversight provided by telegraphy eliminated autonomy, or even disobedience, on the part of diplomats. Some battled to maintain their freedom of action, much like other workers who have resisted technologies that regimented their activities. Second, diplomats performed other functions—such as collecting information, serving as scapegoats, or contextualizing diplomatic messages—that did not require them to act on their own initiative and were therefore unaffected by the use of telegraphy.

Part II opens with the *Trent* affair, an incident that occurred in 1861 when diplomats were just becoming aware of the possibilities and problems posed by the telegraph. Many contemporaries noted that telegraphy had not been used during the *Trent* crisis and speculated about whether the incident might have turned out differently had events occurred at a faster pace. This case study leads to a second question: Did crises become easier to resolve as a result of the speed of the telegraph? After considering the extent to which telegraphy actually accelerated the pace of diplomacy, I examine how faster communication influenced the course of international crises: first, by enticing statesmen to make decisions at moments when public opinion was excited; second, by increasing the strains upon human beings and institutions. I also examine the clash of cultures that occurred when the aristocratic, leisurely world of diplomacy encountered the rapid, industrious world of telegraphy.

Part III presents the tale of the Zimmermann cablegram, which the German government sent—and the British intercepted—in 1917 during the First World War. The incident raises a question: How did the characteristics of telegraphy as a medium affect diplomacy? It also leads to a consideration of several technical aspects of telegraphy. The susceptibility of telegraphic technology to espionage resulted in the creation of new codes, which increased the amount of work required of foreign policy bureaucracies. Telegraph services were also expensive, depending on costly infrastructures and skilled operators. Foreign ministries responded to higher expenses with various cost-cutting measures, such as encouraging concision, which produced changes in the language of diplomacy. Finally, telegraphy was a medium prone to garbling. Slightly distorted telegrams could have disproportionately serious consequences because of the complexity and interconnectedness of the international political system. All these attributes of the telegraph required adjustment on the part of foreign ministries.

Historical Context

When contemplating the rise of telegraphic diplomacy, it is useful to think in terms of three chronological eras: the period before 1851; the period between 1851 and 1918; and the period after 1918. In focusing upon the integration of the telegraph into diplomatic practice, I concentrate mainly upon the second era and, for purposes of comparison, the years preceding it. But first I shall give an overview of all three eras in order to provide a general understanding of the rise and fall of telegraphic diplomacy.

The first era encompasses the years before electric telegraphy began to affect international relations. This era of pre-telegraphic diplomacy ended in Europe by the 1850s, when the capitals of the great powers became part of a single electric communications grid.[4] During the early nineteenth century, navies had moved only as quickly as the wind allowed. Armies traveled no faster than during the Roman Empire. Without instantaneous communication, major battles were sometimes fought after peace had been declared, as when Andrew Jackson's forces defeated the British in the battle of New Orleans two weeks after the official conclusion of the War of 1812. The relatively slow pace of communication before the invention of the electric telegraph often led governments—faced with the need to take quick action—to invest their diplomats with considerable autonomy. The era's dominant political system was based on resurrecting the ideal of absolutist monarchy. During the French Revolution, European potentates had seemed on the verge of collapse. Now, in the wake of the counterrevolutionary restoration of 1815, they oversaw political systems that appeared to have defeated demands for democracy and national unity.

Below the surface, however, social, economic, demographic, and technological changes were undermining the foundations of Europe's apparently stable international system. These changes had become unmistakable by 1848, when they contributed to revolution and political ferment across Europe. During the early nineteenth century the industrial revolution gained momentum in the most advanced economies. Steam engines transformed industry and transportation. Between 1830 and 1851 the combined length of railway line in the United States, France, Great

Britain, and the areas that later became Germany increased from 225 to 37,155 kilometers. The surge in economic activity paralleled rapid population growth. The combined populations of Austria, France, and England/Wales increased from 56 million in 1821 to 71 million in 1851. The population of what would later be Germany increased from 22 million in 1816 to 33 million in 1852.[5]

Many of the industries of the time—especially railroads—fostered precision, greater productivity, and the efficient use of time. In the more clock-oriented society that resulted, many people wanted faster access to information. Financial speculators and journalists generally led the way, as they were the groups most willing to pay for the latest intelligence. Entrepreneurs developed various methods for faster transmission of news. The most elaborate required state sponsorship and served political purposes. During the 1790s governments installed ingenious systems of optical telegraphy. These systems employed chains of stations, each station separated from its closest neighbor by a few miles, to pass visual signals rapidly across great distances (so long as visibility was good). The French semaphore system, designed by Claude Chappe, was the most extensive of these, and helped extend the power of the French state. Improved printing technologies also met the demand for information, and newspaper readership grew. Homing pigeons came into greater use; in 1846 there were 25,000 in Antwerp.[6] Ever increasing demand for information encouraged inventors to experiment with faster media. As of the 1830s, scientists in a number of countries were tinkering with machines that conveyed information over wires via electric impulses. The invention of a commercially viable system of electric telegraphy was just a matter of time. Meanwhile, however, diplomats conducted their activities much as they had for centuries.

The 1850s marked the beginning of a second era—lasting until the First World War—in which governments adopted telegraphy and grappled with the advantages and disadvantages that it offered them in the conduct of foreign policy. This was the golden age of the telegraph, a time before rival technologies could match its capabilities.[7] Two Englishmen, William Cooke and Charles Wheatstone, produced a railway telegraph line in 1837. In 1844 Samuel Morse oversaw the completion of a public telegraph connection between Washington, D.C., and Baltimore. Jacob and John Brett laid a submarine (underwater) telegraph cable between Britain and France in 1851. Cyrus Field directed the herculean effort required to lay an Atlantic cable. After many failed attempts, the Atlantic Telegraph Company laid such a cable in 1858, but it never worked well and soon failed completely. Not until 1866 was telegraphic communication reestablished between Europe and America, too late to influence the diplomacy of the U.S. Civil War. As a result of such technological feats, telegraphy soon became a tool available to the makers of foreign policy.

Some expressed high hopes for the invention. Emperor Napoleon III told an American diplomat that in the age of "the telegraph . . . any misunderstanding . . . might be readily rectified." Telegraphy did promote cooperation between countries. The most obvious example was the International Telegraph Union, one of the earliest international organizations. Yet electric telegraphy did not bring the universal harmony that some predicted.[8] Certainly it promoted international commerce and cross-cultural communication. But it simultaneously challenged internationalist aspirations by serving xenophobic elites engaged in nation building. These elites aspired to create communication systems that would promote economic and cultural integration within their countries and facilitate imperialist conquests beyond their national borders. Likewise, they sought to stoke the fires of nationalism through the creation of government-influenced telegraphic news services.[9] In these ways, autocratic rulers used the telegraph to co-opt the nationalist energies so apparent in 1848, redirecting them away from liberal reform and toward patriotic militarism.

The Near Eastern dispute of 1853 was the first international crisis to occur after the electric telegraph had connected most major European capitals. Governments did not successfully manage the crisis, which led to the Crimean War and the most significant bloodletting between European powers in four decades. The war proved traumatic for both Britain and Russia, two pillars of the post-1815 European political system. Thereafter, the former entered a period of isolationism while the latter pursued a revisionist course aimed at disrupting the status quo. The changed policies of these two countries contributed to a decline in support for the existing European order. The results included several wars among the three remaining powers—France, Austria-Hungary, and Prussia—and the formation of Italy and Germany.

These events spawned the most important changes in the European states system since the defeat of Napoleon a half-century earlier. The crises of this era demonstrated telegraphy's ability to accelerate international affairs. One of the most significant changes in European diplomacy was the rapid speed at which international disputes, most notably the 1870 feud between France and Prussia, could escalate. Symbolically, it seems appropriate that the Franco-Prussian War was the first major conflict whose origins were popularly associated with a telegram.[10] Such crises demonstrated significant changes in military technology. Many observers believed that Prussia owed its rapid military victories over Austria and France to its successful adoption of new technologies—especially the railroad and the breechloading rifle. The general lesson drawn from the Prussian example was that victory in major wars depended upon continuous readiness, elaborate mobilization plans, and large, preexisting forces. Governments observed that disputes could rapidly become war, and that the timing of crises could determine the military victor. As a result, diplomacy took a backseat to military concerns. During crises, even pacifically inclined governments felt pressure to pursue hasty, inflexible initiatives that made military sense but undermined efforts to resolve international disagreements through diplomacy rather than force. Alliances took on a more rigid cast as governments decided that they needed to make preparations for war well before crises began. Once danger appeared, it would already be too late. Enduring and complex alliances became a noteworthy characteristic of European politics during the era before the First World War.[11]

Despite the wars that occurred between 1854 and 1871, European interstate relations maintained a high degree of continuity. The list of European great powers in 1914 remained virtually identical to what it had been in 1815. Furthermore, although there was some growth in the size, as well as elaboration in the structure, of diplomatic services and foreign ministries, they changed little overall. This comfortable stability did not go unchallenged. Some contemporaries, particularly businessmen, called for reform because they felt foreign ministries were too exclusive and too disdainful of commerce. Moreover, the fairly constant membership in the club of European great powers obscured shifts in economic and military strength. Equally misleading was the global preeminence that these powers displayed while subjugating much of Asia and

Africa in a scramble for colonies. Powerful economic, social, and technological forces at work in the world threatened the Eurocentric edifice of international affairs. These forces were best illustrated by the rise of extra-European great powers, most notably the United States and Japan.

Telegraphy retained its primacy as a medium of long-distance communication. It symbolized an imperial age. Diplomats referred to the dispatches they sent electrically between continents as "cables," after the submerged wires that carried the messages. Submarine telegraphy was a technology intimately associated with Victorian England's international ascendancy. Other powers followed Britain's lead, building their own cable networks as they carved out and consolidated empires. Yet the dominance of telegraphy as a medium of rapid international communication—like the hegemony of the European great powers—was more fragile than it seemed. Competing technologies appeared on the horizon. In 1876 Alexander Graham Bell patented his version of the telephone. In 1901 Guglielmo Marconi announced that he had established radio communication between Europe and America. But neither device was yet ready to supplant the telegraph.

In the third era, telegraphy had become a normal medium of diplomatic communication, no longer novel or exciting. One cannot date this era precisely since different governments adopted telegraphy at different paces. But all major foreign ministries used telegraphy with regularity by the time of the First World War. The war required governments to make more extensive use of the technology than they had in the past. As a result, they became more adept at using it. By 1918 governments had made progress against many of the problems associated with sending a telegram—such as garbling and cost. Yet such difficulties remained an inherent element of the technology, and in fact the war had made some problems, such as the danger of telegraphic espionage, more critical than ever.

Three-quarters of a century earlier the electric telegraph had seemed a source of dynamism and innovation. By 1918, however, another communication technology, radio, produced far more enthusiasm than the telegraph.[12] Even for diplomats, generally rather stodgy in their customs, telegrams had lost much of their novelty. Diplomatic habituation to the telegraph was beneficial, for much else was changing on the landscape of international relations. The First World War had produced chaos, destruction, and reform. The pace of reform, even sometimes revolution, in political institutions seemed finally to catch up to earlier changes in the social, economic, and technological realms.

Four major empires collapsed—those of Russia, Germany, Austria-Hungary, and the Ottomans. The regimes of the European "victors"—France, Britain, and Italy—were left shaken. The postwar settlement recognized the influence of non-European countries such as the United States and Japan. In response to the war, the League of Nations came into existence. In much of the industrialized world, popular participation in politics increased. Women received the vote and socialist parties helped form governments. Reformers sought to democratize foreign policy by attacking secret diplomacy and forcing foreign ministries to look beyond the aristocracy for new recruits. Revolutionary ideologies, such as Leninism or fascism, triumphed in Russia, Italy, and Germany.

Leaders extended their influence through the media of movies and radio, allowing them to circumvent their own bureaucracies and appeal directly to the citizenry. The Soviet Union, which had difficulty producing the vacuum tubes used in radios, installed loudspeakers throughout the country to indoctrinate the masses. To an unprecedented extent, technologies of mass communication allowed great numbers of people to feel they possessed a personal connection to figures of world historical importance, such as Roosevelt, Hitler, and Churchill. Charles de Gaulle used radio broadcasts to invent himself as a leader of the French people during the German occupation. With new sources of authority, leaders felt less bound by tradition. They also—to the revulsion of most diplomats—felt less bound by international agreements.[13]

By the beginning of this third era, electric telegraphy was a mature technology in slow decline. During the 1920s short-wave radio communication between continents became cheaper than sending cablegrams, though underwater cables continued to provide a more reliable medium of communication that was less susceptible to atmospheric disturbance. Improvements in the telephone and in mail service cut into the telegraph business; during the 1920s and 1930s the number of telegrams sent in the United States increased only haltingly, while in Britain, France, and Germany use of the telegraph actually declined. By July 1928 Britain's postmaster general noted that every government in Europe ran a deficit in its domestic telegraph service.[14] Technical improvements continued to occur, but they tended to be incremental, such as better means of amplifying electric signals that improved the speed and accuracy of long-distance telegraphy. In terms of labor history, the most significant innovation was the adoption of cost-effective teleprinters (which automatically dispatched outgoing messages and printed incoming ones) during the 1920s. These devices, by drastically reducing the skill required of telegraphers, weakened the bargaining power of skilled Morse operators and reduced costs in what had once been a labor-intensive industry. Whereas Morse operators had been predominately male, the less well paid teletypists were predominately female. During the Great Depression, when companies felt great pressure to increase productivity, most remaining Morse operators retired, took new jobs, or were laid off.[15]

From the standpoint of scientists and engineers, telegraphy was no longer an exciting technology. The action was elsewhere. New media of communication—telephone, radio, and the airplane—increasingly replaced the telegraph. In the eyes of diplomats, however, these new technologies possessed drawbacks that made them unsuitable for the conduct of international relations. Telephone calls, unless they were laboriously transcribed, left no written record. Even when calls were transcribed, speech patterns and vocal affect were difficult to capture on paper. Radio signals were even more vulnerable to espionage than were international telegrams. Air travel facilitated summit conferences, where world leaders might escape from the control of their bureaucratic handlers; in the opinion of diplomats, this was a recipe for disaster. Airplane diplomacy reached its nadir during the Munich crisis, when Neville Chamberlain, a neophyte in matters of foreign policy, flew to Germany to negotiate directly with Hitler. For such reasons, added to their fear of becoming redundant, diplomats clung to telegraphic communication at a time when the rest of society increasingly chose to employ other media. A Belgian politician captured diplomats' suspicion of new technology when he described their opposition to U.S. President Herbert Hoover's use of a radiotelephone during the 1931 international financial crisis: "Diplomats feel they have outlived their usefulness when the head [*sic*] of States can discuss matters almost face to face."[16]

During the Second World War the insufficiency of the German and Japanese cable networks forced those countries to rely heavily upon much less secure radio communication. Great Britain and the United States resolutely exploited this weakness, as they broke the codes of their enemies, anticipated their actions, and, with the aid of allies, defeated them.[17] Out of this cataclysm, two rival "superpowers" emerged, the United States and the Soviet Union. In international politics, the most important event of the late twentieth century was the collapse of the Soviet Union and the end of the Cold War.

The causes of the Soviet decline are complex, but one important factor had to do with the dilemmas it faced while deciding whether to embrace new information technologies. When the USSR was established, in the early twentieth century, technologies such as telegraphy and railroads had produced great advances in communication and transportation in many parts of the world. When combined with other innovations and the harnessing of electricity, they allowed the development of heavy industry (such as steel making) on a massive scale. The Soviet Union relied upon its enormous endowments of labor and natural resources (especially cheap energy) to industrialize quickly and create economies of scale.[18] Soviet factories churned out vast quantities of mass-produced items, some of them of reasonably good quality (such as the outstanding T-34 tank used against Germany during the Second World War). Politically, the Soviet government found that it could employ many information technologies—printing presses, film, and later radio and television—to broadcast messages that bolstered its power.[19] In Soviet hands, the state telegraph agency lent itself to top-down, centralized control as it swiftly conveyed directives and permitted closer oversight of subordinates.

Even when the Soviet regime was at its zenith, however, communication technologies threatened a state that sought total control over the lives of its citizens. Because of the government's ambitions, the telephone was never allowed to achieve its full potential in the USSR. Few could obtain telephones, and those who did discovered that the system was of poor quality and often malfunctioned. Consumers had no expectation of privacy, and were consequently guarded in their conversations. Individuals who were not members of the political elite could only call people whose numbers they already knew because no "phone book" was publicly available.[20] Likewise, the Soviet government severely restricted the use of photocopy machines. Computers posed even more stark trade-offs between efficiency and the regime's social control. One scholar estimated that by mid-1988 the United States, despite a smaller population, possessed 30–40 million personal computers, in comparison with 100,000–150,000 in the USSR. Did these numbers indicate the Soviet regime's strength or weakness? Certainly the relative scarcity of publicly accessible computers helped protect the Communist Party's dominance. Yet Soviet leaders knowledgeable about the development of Western economies believed that computers must play a vital part in the future development of their own country.[21] Moreover, foreign policy intellectuals and military leaders realized that the Soviet armed forces would become second rate without access to the latest information technologies. The spread of these technologies throughout society, however, weakened the grip of the regime. Indeed, when the Soviet government did loosen its control in order to exploit the benefits of an information-based economy, it eventually faced a choice between abandoning its attempts at economic modernization and letting its authority crumble. It chose the latter course, and the United States became the world's sole superpower.[22]

Meanwhile, foreign ministries encountered new technologies of communication. The laying of the first long-distance submarine telephone cables in 1956 (which made all existing telegraph cables outdated) and the launching of the first communication satellite in 1962 pushed telegraphy toward obsolescence in the realms of business and journalism. Increasingly, the electric impulses that composed telegrams were sent over telephone lines, which had greater carrying capacity and were more cost-efficient than telegraph wires. Foreign ministries, however, continued sending "cables" despite the diminished status of this declining technology.

During the late twentieth century optical telegraphy reversed its earlier defeat by electric telegraphy. Fiber optic cables transmitting light proved more efficient than copper wires conveying electricity. The use of light as a communication medium furnished the immense transmitting capacities conducive to the World Wide Web. The Internet opened to commercial use in 1992 and reached about 500 million people a decade later. By the early twenty-first century about half the inhabitants of the developed world were online, where tens of millions from the less developed countries joined them.[23]

At the end of the twentieth century diplomats still sent "cables" and "telegrams," but much was changing. Diplomatic "cables" no longer necessarily traveled by cable, as foreign ministries had begun to use wireless communication (especially radio, satellite, and microwave transmission) to supplement wired networks. The meaning of the word "telegram" likewise became blurred. Fax machines and e-mail, like the telegraph, converted readable text into electric signals and back again. Increasingly, "telegrams" and "cables" were defined by the clearances (authorizations) required to send them, and by their formality—especially the way they were handled and recorded—rather than by the medium through which they traveled. "Telegraphy" as a distinct technical system disappeared. Instead, the term designated particular bureaucratic procedures associated with a now antiquated technology. And even the term began to fade. In 1998 Britain's Foreign and Commonwealth Office announced its intention to end use of telegraphy by the year 2000. The U.S. State Department, which entered the twenty-first century with the "cable" still its most formal medium of diplomatic reporting, likewise prepared to move beyond a system that it now found "obsolete." The age of diplomatic telegraphy—which had lasted about a century and a half—was ending.[24]

Control I

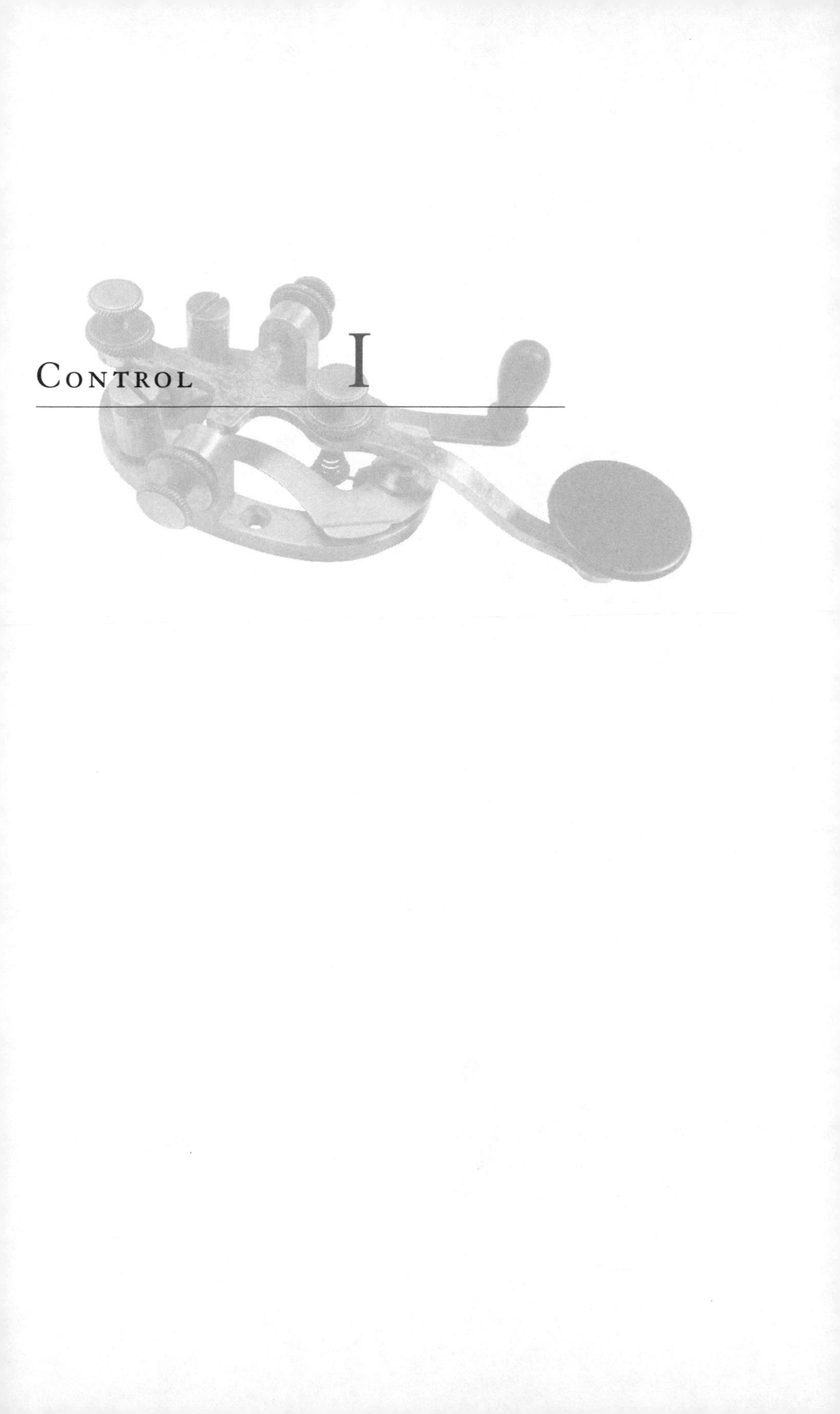

The Anglo-American Crisis of 1812

1

In mid-June 1812 the U.S. Senate faced a momentous decision: whether or not to declare war on Great Britain. Venomous relations had existed for years between the United States and its former colonial ruler. Among the explanations commonly given for American hostility toward Britain at this time are the effects of economic depression, vexation over alleged British encouragement of Indian raids, land hunger (for British North America and Spanish Florida), and the explosive interaction of British arrogance with America's prickly nationalism. Not least among the sources of tension, "impressment," the British practice of seizing alleged deserters from the Royal Navy serving as sailors on American ships, outraged the American public.

Despite the fury over impressment, a strong case can be made that the most important of the many sources of conflict between the two nations was the "Orders in Council."[1] The Orders were a long, complex, and oft-revised series of regulations that by 1812 required Americans trading with continental Europe to receive British permission and pay a fee. This policy grew out of the war effort against France. The Orders forced neutral trade with Europe to pass through Britain. They undermined French power and provided a mechanism to prevent war-profiteering foreigners from gaining mercantile advantages lost to Britain because of Napoleon's policies. In effect, however, the Orders favored British merchants over their rivals from the United States and struck many Americans as an unfair form of protectionism aimed at destroying the young republic's mercantile prosperity. The Orders possessed symbolic importance as well: they seemed a humiliating sign that, a generation after the Revolution, the United States retained colonial status in Britain's eyes. President James Madison, speaking during the war, insisted that meek acceptance of such rapacious maritime policies would have been an admission that "the American people were not an independent people but colonists and vassals." Moreover, America's weakness against British naval aggression reinforced the fear that nations with a republican form of government were incapable of defending their rights in the international arena.[2]

On 1 June 1812 Madison issued a presidential war message. Three days later the House of Representatives affirmed this declaration by voting 79–49 in favor of war. The last barrier to open hostilities seemed to be the Senate, where measures to launch only limited naval hostilities against Britain, to postpone action entirely, and to declare a triangular war against both Britain and France lost by narrow margins. Any of these actions would have slowed the pace of events and provided additional time for diplomacy before the United States irrevocably committed itself to war. Most senators opposed an all-out war with Britain, but many feared the political ramifications of a retreat after so much bellicose rhetoric. Finally, on 17 June 1812, the Senate voted 19–13 to declare war.[3]

According to the British minister in Washington, Augustus Foster, the vote would have been closer but for "two or three" senators who, seeing the inevitability of war, decided "to swell the majority." To his disappointment, Foster found that even his efforts to win over an alcoholic senator by supplying him with liberal quantities of liquor had failed. Ensuring that the senator was "drunk every day . . . was no difficult matter," but Foster's tactic did not prevent the Virginian from voting for war.[4] Nonetheless, the Senate's vote would prove to be the closest of any declaration of war in American history, following a debate unparalleled in its length. At three o'clock on the afternoon of 18 June, President Madison signed the declaration. The United States was at war with Great Britain.

Meanwhile, a series of dramatic events had occurred across the Atlantic. On 11 May 1812 a lunatic murdered Spencer Perceval, the prime minister of Britain and the author of the Orders in Council. This incident, although unrelated to the controversy over the Orders, seemed to symbolize growing public dissatisfaction with Britain's commercial policy toward the United States. By mid-1812 many in Britain believed that the Orders had worsened, or perhaps even caused, an economic depression. British industrialists criticized the Royal Navy's treatment of U.S. merchants, arguing that American anger had produced a damaging boycott of British goods (the Non-Intercourse Bill passed by Congress). Opponents of the Orders also attributed widespread rioting to their economic effects. But other pressing issues and the difficulty of forming a new government after Perceval's death distracted attention from the controversy over the Orders. The issue was allowed to drift.[5]

Finally, on 16 June, one day before the U.S. Senate passed the war bill and two days before Madison signed it, the parliamentary opposition, after weeks of testimony on the harmful consequences of the Orders, introduced a motion calling for their repeal. Viscount Castlereagh, the British foreign secretary, responded on behalf of the government. During the debate he announced "that a proposition should be made to the American government to suspend immediately the Orders in Council, on condition that they would suspend their Non-Intercourse Act." The revocation of the Orders brought rejoicing throughout England; it seemed that the American market would soon be open again to British goods. Instead, to their horror, British industrialists and workers learned six weeks after Castlereagh's announcement that the repeal had been too late; the United States had declared war. Ships carrying the British concession had crossed others carrying news of war.[6]

Telecommunications and the Origins of the War

In 1812 the relationship between electricity and magnetism was not well understood. The construction of a successful transatlantic telegraph cable lay half a century in the future. On average, ships required a month to sail from England to the United States, although weather conditions produced considerable variation in travel times.[7] Aware of these facts, many scholars have declared that a telegraph cable across the Atlantic would have allowed Americans to learn of Castlereagh's 16 June revocation of the Orders in time to prevent the War of 1812.[8]

One should treat such counterfactual statements with skepticism since causal factors are often interdependent—and in fact the sluggish pace of ship-borne communications had impeded the outbreak of an Anglo-American war on at least two previous occasions.[9] Nonetheless, this view has merit. If the Senate had known of Castlereagh's statement, it would not have voted for war. Likewise, Madison would not have signed the declaration had he been aware of the shift in British policy. Speaking to a historian two decades later, the former president contended that if Britain had rescinded the Orders a few weeks earlier, "war would not have been at that time declared, nor is it probable that it would have followed, because there was every prospect that the affair of impressment + other grievances might have been reconciled after the repeal of the obnoxious orders in council."[10]

Other well-placed individuals agreed with Madison. James Monroe, the U.S. secretary of state, told the British chargé d'affaires that the maintenance of the Orders "was the only material point of difference between the 2 countries. Other matters might be accommodated or made matter of negotiation." Augustus Foster suggested that news of the repeal would have stayed a declaration of war, as did the U.S. representative in London, Jonathan Russell. Far away, John Quincy Adams, the American minister to Russia, wrote to his mother: "I then flattered myself that the revocation of the British Orders in Council . . . would be known in the United States in season to prevent the war which I knew would otherwise be unavoidable. In this hope I have been disappointed . . . I lament the declaration of war as an event which in the actual state of things when it passed was altogether unnecessary, the greatest and only insuperable causes for it having been removed."[11]

Once America's political leaders had declared war, they persevered in spite of this major British concession because to revoke the declaration would have made them and their country look ridiculous, especially since there were other sources of friction in the Anglo-American relationship. Nonetheless, the origins of this war force one to reflect upon the tragically slow pace of transatlantic diplomacy in June 1812.

Diplomatic Reporting and Deceit

The Anglo-American crisis of 1812 is a remarkable example of the potentially decisive role played by technologies of diplomatic communication. But do the causes of this war illustrate any general phenomena about long-distance communication relevant to the study of other international crises? Two issues suggest themselves. The first relates to the disparity visible in this instance between the relative speeds of international and domestic politics. The second concerns the tendency of statesmen to behave deceptively. It reminds us that communication media lend themselves in varying degrees to the centralization and control of information, with implications for incentives to engage in deceitful diplomacy. It also illustrates the costs of secrecy.

First, the crisis of June 1812 reveals an asymmetry between the speed of political decisionmaking and the speed of transatlantic communication. Compared with the torpid pace of an Atlantic crossing, the British government rescinded the Orders in Council with a celerity that caught the United States off guard. Likewise, British observers were surprised by what seemed to be a sudden movement toward war in Washington after a long period of vacillation. This asymmetry resulted from the close relationship between distance and communication speed that existed before the mid-nineteenth century.[12]

Electric telecommunications across the Atlantic would have allowed British and American makers of foreign policy to respond in a more timely fashion to rapidly changing events overseas and helped them avert inappropriate but irrevocable actions such as the American declaration of war. In London, the *Times* commented: "The haste of Congress to declare war against us, has outrun the slowness of the arrivals from England."[13] Had a cable existed at this time, domestic politics would not have outrun diplomacy so easily.

This lack of speed in long-distance communication was not a new problem. The tempo of domestic politics had often exceeded that of international politics in the past. Why were the consequences so disastrous in this instance? The answer has to do with the solutions that statesmen found to the difficulties of coordinating foreign policy across great distances in the age before electric communication. In general, foreign ministries compensated for slow communication by giving diplomats greater autonomy (see Chapter 2). Issues of war and peace, however, exacerbated coordination problems; political leaders did not wish to delegate control over such important matters to agents in distant countries. Governments struggled to make decisions that would still be appropriate weeks later, when their dispatches finally reached overseas destinations. To meet this challenge, statesmen relied on their diplomats to act as "early warning systems" sending back insightful and prescient reports. Given the state of communication, misperceptions were difficult to correct and might derail diplomacy for a protracted period; Augustus Foster told President Madison that "a mistake in knowing the views of each other, would create three months delay." Unfortunately, neither the British nor the American diplomats in one another's capitals produced the high-quality reports that were needed in 1812. The failures of diplomatic agents interacted with the problems caused by slow transatlantic communication, producing a war that neither country desired.[14]

Personnel changes impaired the ability of the U.S. government to collect accurate intelligence about British politics. On 28 February 1811 William Pinkney, the U.S. minister to Britain, made his farewell appearance as minister at the Court of St. James's. Madison did not replace this able diplomat with someone of the same caliber and rank, apparently because he wished to express American displeasure toward Britain. Jonathan Russell became the head of the American mission, but he was only a chargé d'affaires. Secretary of State Monroe instructed Russell "to communicate the most full and correct information" on whether Britain would maintain the Orders in Council. But Russell's lack of ministerial rank meant that he could not move effectively within the higher levels of the British government.[15] Moreover, Russell, known for a cantankerous demeanor that made him an improbable diplomat, had strong incentives to return to America. His financial circumstances were in serious decline, and his wife had died some months earlier, leaving four children in America who needed care. Preoccupied, isolated, and unhappy, Russell misread the British political scene. Petitions against the Orders poured in from industrial towns while opposition MPs forcefully condemned them in Parliament. Nonetheless, Russell sent reports to the State Department asserting that there was little hope for a repeal of the Orders and that the United States must resort to force to defend its rights.[16]

On 8 June, about a week before parliamentary sentiment forced the revocation of the Orders, Russell wrote to Monroe that he wished to leave his post in case of war and that, if war did not come, "I hope it will not be required of me to continue here . . . I must frankly avow that my nerves are [not] proof to the contempt and contumely to which I should be exposed here should [our] government, after its repeated declarations

[and] preparation, surrender without a blow. Should such, indeed, be our humiliating course . . . the sound of our instruments of war will have served only to attract the attention of the world + to render our dastardly flight more notorious + more disgraceful." Not until his dispatch of 9 May, when it was already too late to reach Washington before the American declaration of war, did Russell first indicate that the British government might conceivably revoke the Orders. But even at this late date he suggested that "an immediate removal" of the Orders was unlikely and, should it occur, would "probably be productive of very little either for the satisfaction of our honor or the promotion of our interests."[17] At the beginning of July 1812, before officially hearing of the U.S. declaration of war, he made plans to abandon his post and return home without permission. In sum, the United States was represented in London by a diplomat desirous of war and eager to leave his post as soon as possible.[18] Because of his reporting, the British government's repeal of the Orders took the United States by surprise.

The lack of effective American diplomatic representation in London made the role of British diplomats in the United States all the more important. Unfortunately, matters were scarcely better on the western side of the Atlantic. The top post of the British legation was briefly held by Francis James Jackson, who reached Washington in September 1809. Jackson's very appointment insulted American sensibilities as his chief claim to renown was the delivery of an ultimatum to Denmark before the British fleet bombarded Copenhagen in 1807. Once in Washington, Jackson made himself so obnoxious that within months President Madison asked for his recall and Congress voted to condemn him. Instead of leaving when his mission failed, he stayed on almost a year, socializing with the opposition Federalists, subsidizing their attacks on the Madison administration, and further enraging the Republicans who controlled the American government.

Pinkney's departure from London induced the British government to send Augustus Foster to Washington as the new British minister. Foster arrived in the United States on 29 June 1811, finally filling a position that had been vacant for about a year. Foster was amiable, intelligent, and knowledgeable about the United States, having served as secretary of the legation in Washington from 1804 to 1808. Unfortunately, he had a low regard for the country to which he was appointed and accepted the posi-

Augustus Foster, British minister to the United States in 1811 and 1812.

Library of Congress, Prints and Photographs Division, LC-USZ62-10568. Engraving from *The Two Duchesses,* ed. Vere Foster (London: Blackie & Son, Ltd., 1898)

tion only because the diplomatic posts of Europe were closed to him by Napoleon's conquests. When Foster's mother heard about his nomination as minister to the United States, she urged him to accept in spite of his "dislike [for] that country."[19]

During Foster's earlier residency in America, the quality of American society had left him unimpressed. "Pray don't marry an American," his mother needlessly warned, but the wellborn young man had already concluded that American women were "in general a spying, inquisitive, vulgar, and most ignorant race." As for their male counterparts, "from the Province of Maine to the borders of Florida you would not find 30 men of Truth, Honour, or Integrity." With a mixture of sadness and amazement, Foster concluded that "the scum of every nation on earth is the active population here." The conduct of an American congressman/tavern keeper, who, dismayed at the prospect of a long trek to a cold outhouse, used Foster's fireplace as a urinal, only confirmed the diplomat's prejudices.[20]

Foster's anti-Americanism made even more onerous the duty of representing a country that many Americans despised. In these circumstances, it was understandable but unfortunate that while he went out of his way to meet political figures of all stripes, he, like Jackson before him, gravitated to the Anglophilic and genteel members of the Federalist party. The information he gathered from them contributed to his tendency to discount warnings from the Madison administration about the possibility of war. The Federalists, behaving almost treasonously, repeatedly told Foster that Republican talk of war was nothing more than political posturing.[21] While most of the Federalists were probably expressing their true opinion, they also stood to gain from discouraging British concessions to the United States at a time when the Republicans held power. Madison aggravated this situation, perpetuating Foster's ignorance and confusion by refusing to clearly indicate the circumstances under which the United States would be likely to carry out its vague threats of war.

Further complicating matters, the woman Foster wished to marry was still in Britain (she had recently met Lord Byron, whom she eventually married). Foster had considered refusing the post in Washington so that he could continue his courtship. Once in America, he proclaimed his refusal "to add to the number of diplomatic old bachelors." He longed "most anxiously to get back [and resolve the relationship]. No Minister ever had such temptations to break up a negotiation. I would give the world to go back for six months."[22] On 24 April 1812 he indeed asked for a six-month leave of absence "to arrange my private affairs." He suggested that his absence from Washington would have little effect and declared, "My hopes of a favourable issue to the present difficulties with [the United States] continue to increase." This was less than six weeks before Madison sent his war message to Congress. After the outbreak of war, Foster would leave for Halifax and then return to Britain, penning a long dispatch in an attempt to justify his decision to return home on his own initiative.[23]

Not until 15 June 1812 did Foster's dispatches warn of a full-blown

Anglo-American war. Madison signed the congressional declaration of war a mere three days later.[24] Foster, like Russell in London, failed to fulfill his role as an information conduit between Britain and the United States. The resulting intelligence vacuum left his country unable to foresee the American war declaration. Plagued by this inadequate reporting, Britain's foreign secretary, Lord Castlereagh, did not grasp the perilous state of Anglo-American relations. When Castlereagh finally announced the suspension of the Orders in June 1812, he indicated that he thought war unlikely and "lamented" the Opposition's excessively "hasty" demands for repeal.[25]

Why was Foster so slow to grasp the likelihood of war? Much of his difficulty in accurately evaluating the American political scene resulted from his reliance on Federalist sources. His lack of sympathy for American life and his preoccupation with his romantic difficulties exacerbated the situation. Moreover, the analysis of U.S. foreign policy in 1812 was not a simple task. The factiousness of Congress and the secrecy of the Madison administration obscured the course of political life in Washington.

This last point, the behavior of the Madison administration, raises a second issue dramatized by the culmination of the Anglo-American crisis in 1812: the use of diplomatic communication to deceive other nations. President Madison had decided by 1811 that the United States must go to war unless Britain ended its application of the Orders to American shipping.[26] This resolution did not strengthen American diplomacy because the Madison administration, while engaging in belligerent rhetoric that alarmed some British observers, told Foster that the United States was not about to go to war. For example, in late 1811, when Foster asked Monroe to comment upon a summary of the American position that the British envoy intended to submit to his superiors in London, the U.S. secretary of state asked for changes to remove the impression that the United States was menacing Britain.[27]

A U.S. embargo on trade with foreign countries, signed into law on 4 April 1812, typified the administration's mixed messages. In secret debates preceding the embargo, the administration portrayed it as a policy leading to war. The president also depicted the embargo in this light in a letter to Thomas Jefferson on 3 April. But the very day he wrote to Jefferson, Madison gave the opposite message to Foster, telling him that "Embargo is not war."[28] Similarly, the secretary of state obstructed Foster's effort to learn the true purpose behind the embargo bill: Monroe "deprecated its being considered as a war measure." As Foster later noted, "The system of . . . holding a double language is indeed now so much the practice of this Government that to judge of what they mean by what they say is utterly impracticable." Faced with such tactics, the British minister complained of the American government's use of "trick, falsehood and artifice."[29]

Why did the Madison administration undercut its own diplomacy by refusing to send unambiguous warnings to Britain? Madison had apparently decided that diplomacy should be sacrificed to military considerations. Forgoing diplomacy did not seem a great loss since the president, influenced by years of barren Anglo-American negotiations and Russell's recent reports, saw little hope of significant British concessions. At the same time, the administration feared that revealing U.S. military intentions to Britain, while it would have made American diplomacy more credible, would also have deprived the United States of the advantages of a surprise attack and, indeed, might have provoked Britain into launching a first strike of its own.[30]

Before discussing the issue of surprise attacks, one must note that neither the United States nor Great Britain could mortally wound the other in 1812. The population growth of the United States, combined with its vast territory and decentralized political and economic structure, made it virtually impervious to British reconquest. Likewise, the unquestioned superiority of the Royal Navy following Nelson's victory at Trafalgar precluded any American threat to Britain. Thus the possibility of a surprise attack did not affect the survival of either country.

Nonetheless, the U.S. government believed that surprise would produce military advantages in 1812, and it feared a preemptive strike by Britain. Americans knew that in 1807 Britain had shelled Copenhagen and seized the Danish fleet out of concern over the possibility of an alliance between France and neutral Denmark. In 1811 Albert Gallatin, the secretary of the treasury, advised Madison to soften an upcoming speech to Congress to avoid conveying the impression that the United States planned to declare war immediately unless Britain repealed the Orders.

If the British decided against repeal despite American threats, Gallatin explained, "they would strike at once." Madison followed this advice, removed menacing phrases, and thereby gave British observers a false sense of security.[31]

Peter B. Porter, the chairman of the House Foreign Affairs Committee, also worried that the United States might provoke a preemptive British strike. On 18 February 1812 he warned: "In the numerous wars in which she [Britain] had been engaged, she had rarely been known to give the second blow. And if we continued to go on preparing for war in the good natured, desultory way we had hitherto pursued . . . we shall receive a stroke which this nation may long have occasion to lament; and for which, if justly imputable to our tardiness and indecision, we could never be forgiven." Madison feared that if war began while the United States was unprepared, the Royal Navy would lay waste to the American merchant marine. Motivated in large part by this anxiety, the president successfully advocated passage of the embargo bill that preceded hostilities. On 5 May 1812 Monroe wrote to Russell requesting that he "apprize our Merchants and other citizens" around the British Isles of the "probability of a rupture" between the United States and Great Britain. He asked that Russell do so "in a way least calculated to excite alarm or to attract the attention of the British government."[32]

Both House and Senate debated in secret session whether to declare war. This concealment was partly an attempt to keep Foster ignorant and hinder efforts to warn the British government about the imminence of war. After the declaration of war, Foster claimed that the actions of the Madison administration indicated the naval advantages the United States hoped to acquire through surprise: "Altho' the American Minister did not refuse to allow me to dispatch a vessel to England on the event of the War, yet the delays which seemed to have been affectedly made at the Department of State, in sending the necessary orders, and in preparing the Passports, made too evident that they wished to prevent, for some days, the departure of any vessel that might give warning of the state of affairs, to His Majesty's ships . . . which were said to be off Sandy Hook." Foster also noted that "the War Party . . . expected to capture some of our frigates, which were said to be near the Coast, + to prey upon our trade in every direction with impunity." The U.S. government was loath to give up the advantage of surprise and refused Foster's request that the American declaration of war be suspended until news reached England.[33]

All forms of communication lend themselves in varying degrees to the practice of deception. Once Madison had decided upon war, he misled the British government in order to exploit the slowness of transatlantic communication. He knew that if American plans for war could be kept secret from Britain, the British government would not learn of them until many weeks after the U.S. declaration of war. This element of surprise might provide offensive opportunities for American forces against their unprepared British counterparts, but most important, it would protect Americans against a preemptive attack by Britain. Unfortunately, these military considerations hindered American diplomacy and prevented a peaceful resolution to the crisis. The British government, already fighting Napoleon, had no desire for another war in 1812 and might have behaved differently had it received better information about American intentions.[34] For his part, Madison would not have led the United States into war had he better understood political events in Britain. Yet much of his ignorance was his own fault. His administration refused to appoint a new minister to London, deceived Britain's hapless minister to the United States, and restricted intercourse between the United States and Britain through such measures as the embargo. Although diplomatic communication with one's adversaries can be difficult, frustrating, and politically unpopular, Madison's experience reveals the costs that can result when governments pursue a policy of secrecy and pique.

Madison's exploitation of the slowness of communication brings us back to the question of a transatlantic telegraph cable: How would such a cable have affected incentives for deceptive diplomacy in 1812? In some contexts, telegraphy increased inequalities in access to information. In this case, however, a transatlantic cable, by accelerating the exchange of information in Anglo-American relations, would have made each country's foreign policy (as expressed through diplomatic messages, newspaper stories, and financial transactions) more transparent to the other. Faster conveyance of news about military measures, by reducing the length of time in which to wage war against an ignorant enemy, would have diminished the advantage of being first to begin hostilities. As a result, the United States would have had less reason to fear a preemptive attack by Britain, and less incentive to declare war.[35]

This tale suggests two hypotheses that may be usefully applied to diplomacy during other historical epochs. First, when technologies of diplomatic communication tended to produce unequal access to information, they encouraged greater use of deceptive diplomacy. Second, the attributes of individual diplomats tended to be more important during eras of relatively poor communication.[36] The inept reporting of Russell and Foster undermined efforts to resolve Anglo-American tensions peacefully. Decades later, the diplomatic successors to Russell and Foster discovered that electric telegraphy breached their isolation, contesting the independence, influence, and status associated with their profession.

Diplomatic Autonomy and Telecommunications 2

The electric telegraph was fast. Its speed allowed foreign ministries to oversee and direct the activities of diplomats more easily. It also offered a scarce resource. Unequal access to the telegraph produced more hierarchical divisions of labor within foreign policy institutions. These two characteristics, speed and scarcity, diminished the independence of diplomats and furthered the centralization of foreign ministries. But although telegraphy contributed to these broad trends, its effects were complex and variable, differing greatly according to context. Different foreign ministries offered different environments for the reception of the telegraph.

Such differences produced great diversity in the degree of independence that various governments accorded their diplomats. During the late nineteenth and early twentieth centuries, German and Habsburg diplomats tended to demonstrate relatively little autonomy. French, Russian, and U.S. diplomats showed considerably more. British diplomats fell between these two extremes. Diplomats had mixed feelings about the encroaching power of foreign ministries. Many appreciated the increased guidance they received. Others bridled at efforts to centralize control over foreign policy. But contrary to the conventional wisdom—which emphasizes the characteristics of telegraphy that enhanced control over subordinates—certain aspects of telegraphic communication facilitated diplomatic resistance.[1] In particular, the telegraph's cost, lack of security, and tendency to garble messages furnished independently minded diplomats with means to subvert their instructions and maintain their freedom of action.

• • •

Speed was the telegraph's most obvious characteristic. Telegraphy was a technology that offered foreign ministries an unprecedented capability: the power to respond to distant events as they occurred. Contemporaries believed that the telegraph's rapidity annihilated distance. And indeed, telecommunication techniques helped configure notions of space and time. Because the experience of distance is to a large extent temporal, the world seems smaller after improvements in transportation and communication. A former American diplomat asserted that steam power and the telegraph were among the new forces that "have brought foreign countries to our door and have carried us to theirs."[2] Perceptions of space and time, in turn, have shaped the development of diplomatic institutions.

The tension between proximity and distance provides a prism for understanding the history and purpose of diplomacy. A sense of proximity was crucial to the development of regularized diplomacy; propinquity produced the contact and interaction that created an incentive for institutionalized relations between states. The first permanent embassies originated among fifteenth-century Italian city-states as a response to the acceleration of international relations. Those who did not keep continual tabs on dynamic and expansionist neighbors were likely to find their city conquered. A sense of nearness led states to reform their mechanisms for conducting diplomacy.[3]

Yet the perception of distance also shaped the evolution of foreign policy institutions. Some of the most significant functions of diplomats, such as conveying messages and negotiating in absentia, began in response to the inability of the sovereign to rule at home while negotiating at a far-off location. It was distance that imbued diplomats with their functions and powers. Viscount Castlereagh, British foreign secretary in the early nineteenth century, noted, "The remoteness of America from the events passing in Europe mak[es] it of the utmost importance to the cultivation of a good understanding that an accredited minister, fully authorized to act for them in the many delicate cases that necessarily grow out of the present state of Europe and measures adopted by the Belligerents should be resident at this Court."[4] The practice of sending diplomats to reside in foreign capitals arose when states felt close enough to perceive the usefulness of regular communication, but also because states were frustrated by distances that made communication require intermediaries based in foreign countries. Accordingly, telegraphy had somewhat contradictory effects upon diplomacy: by shortening perceived distances, it manifested the need for greater international cooperation (increasing the need for diplomacy); simultaneously, it tended to reduce the autonomy of diplomats (diminishing their importance).

In arenas other than diplomacy, the speed of telegraphy often promoted control and centralization. It allowed leaders to monitor and amend the activities of distant followers in response to the unfolding of events. For example, telegraphy played an important part in the rise of big business during the second half of the nineteenth century.[5] Corporations found that the telegraph facilitated their efforts to construct large, geographically dispersed businesses and to coordinate complex commercial activities across vast distances. In addition, better communication allowed them to internalize within their own organizations many activities that outside agents had previously performed. Although such growth often increased efficiency and pushed down costs for consumers, not everyone had reason to celebrate; many formerly independent businessmen now found themselves forced to find new jobs. For example, telegraphy tended to undermine the role of middlemen in such businesses as the trades in grain, iron, cotton, tin plate, and meat.[6] Telegraphy expedited the efforts of "robber barons" to build empires by cannibalizing small businesses. To the individuals affected, it seemed only fitting that Western Union, the leading telegraph corporation in the United States, was controlled by Jay Gould, probably the most despised American businessman of the 1880s.[7]

Telegraphy offered some institutions greater control over their own distant agents who, though formally subordinate, exercised considerable independence. For example, the telegraph made it possible to exercise military command from afar. The first instance of this phenomenon occurred during the Crimean War, when French and British military officers found themselves bombarded with bewildering telegrams and exasperated by the interference of civilian political leaders. Helmuth von Moltke, the military hero of German unification, introduced field telegraph units into the Prussian army. Although the technology increased his control over subordinates, Moltke worried about the problems that resulted from having a "telegraph wire in the back."[8]

Meanwhile, British and French colonial officials looked hopefully to the telegraph, in the belief that it would allow them to better supervise excessively energetic proconsuls in the tropics. Although these hopes were not entirely fulfilled, many colonial agents, such as Britain's Lord Cromer, affirmed that the connection of England and India by telegraph enabled the colonial secretary to exercise much more power on the periphery than he formerly had. The British writer George Orwell, who spent time as a colonial official in Burma, remarked that telegraphy reduced "the one-time empire builders . . . to the status of clerks, buried deeper and deeper under mounds of paper and red tape."[9]

The telegraph's provision of a scarce resource also affected bureaucratic hierarchy. Whether through high prices or through systems of rationing, access to the telegraph was restricted, particularly in its early years. One might ask whether telegraphy promoted egalitarian or authoritarian modes of interaction.[10] Many aspects of the technology tended to produce centralization and the amassing of authority. Long-distance telegraphy tended to concentrate message traffic along particular lines or cables, producing bottlenecks and choke points. In such circumstances, individuals could monitor and control the sending or reception of messages at their end of the line. John Kenneth Galbraith, as U.S. ambassador to India, used language that he later recommended to his fellow chiefs of mission: "All . . . communications . . . go through me." This would ensure that "you know fully what is going on and can act as necessary against that of which you disapprove, including vetoing the transmission itself." Furthermore, telegraphy—unlike media capable of directly broadcasting to a large audience—operated from point to point. As a result, some people received information before others and became conduits for its diffusion.[11]

The high prices and rationed access characteristic of the telegraph resulted from the fact that telegraph systems required skilled labor and an expensive infrastructure and, in turn, provided a scarce resource. (At first, telegraph lines could carry only one message at a time. Technological improvements later allowed wires to carry multiple messages simultaneously, but the carrying capacity of telegraph wires remained extremely limited.) The high cost of telegrams favored wealthy individuals and institutions, granting them access to the latest information. A businessman noted that "as the line is now worked the small man is placed in an unfavourable position as regards the large man." The limited carrying capacity of telegraph lines allowed those who controlled the lines to ration access to them. For example, domination of the telegraph cables by bellicose, pro-British news sources slanted the information available to Australians at the start of the Boer War. News from a dissenting perspective, such as that of the *Manchester Guardian,* arrived in Australia by boat rather than by wire, with a delay of more than a month.[12]

But telegraphy could promote equality among those with access to the network, especially when this access was equitably apportioned. In Britain telegraphy evened out the distribution of information between newspapers. Before the telegraph, provincial papers had little ability to compete with London's *Times* in the collection of foreign news. After the development of wire services, these papers could run international stories at the same time as the *Times.* In the United States, too, regional newspapers found that they could escape the hegemony of metropolitan dailies by subscribing to a telegraphic news service. With the creation of the Associated Press, U.S. presidents no longer found it essential to control at least one newspaper in the nation's capital. The power of the press had become more widely distributed.[13]

Yet, on the whole, telegraphy promoted hierarchical structures. Foreign ministries offer a good example. The cost, secrecy, and aura of importance surrounding telegrams made foreign policy officials hesitant to devolve authority over processing them to lower-ranking bureaucrats. In 1875 the U.S. secretary of state decreed: "No messages are to be sent by the telegraph except those from the Secretary, or one of the Assistant Secretaries, or the Chief Clerk, or those authorized to be sent by one of the above. Any chief of Bureau or Clerk wishing to transmit any message on public business, by the telegraph can do so by obtaining the endorsement of the chief clerk." Likewise, the British Foreign Office restricted second-division clerks from processing telegrams until 1911. Pressures toward greater hierarchy also affected overseas legations. In 1909 the U.S. Department of State decreed that only the head of mission and first secretary were allowed to handle the main codebook used for decrypting secret telegrams. In France the foreign minister asked heads of mission to provide written justification for each telegram they sent.[14]

In practice, such regulations often proved unworkable. Sustained periods of crisis usually resulted in the relaxation of rules about processing telegrams; for example, the British Colonial Office's restrictions against lower-ranking clerks deciphering telegrams gave way during the Boer War. The laziness or incompetence of senior officials also sometimes undermined the concentration of authority produced by telegraphy. In 1917 the U.S. embassy in Vienna received a highly confidential telegram concerning efforts to detach Austria-Hungary from its alliance with Germany. It was marked "to be deciphered by the Ambassador only," but the ambassador "had not the slightest idea how to use the code" and gave it to one of his subordinates to process. Even when ignored, however, restrictions on who could send and handle telegrams were symptomatic of the manner in which telegraphy contributed to the development of bureaucratic lines of authority.[15]

The telegraph's tendency to strengthen central authority did not operate only within foreign ministry buildings but extended across national borders to reduce the power of distant diplomatic envoys. Although colonial agents complained that the telegraph deprived them of freedom, diplomats suffered an even greater loss of autonomy. Ambassadorships to powerful nations had formerly been the most important posts in the diplomatic service. Diplomats had often been able to shape the content of the instructions sent to them by monopolizing the information reaching their home governments.[16] In the age of the telegraph, however, they found themselves competing with journalists and businessmen to gather and instantaneously transmit the news. The largely unregulated activity of these alternative collectors of information provided home governments with new sources of information to use when formulating foreign policy and overseeing diplomatic envoys. Diplomats found that shaping their own instructions became more difficult under such circumstances. Diplomats stationed in the great powers typically had less autonomy, and less ability to manipulate the information sent back to their home governments, than envoys in less developed nations (with less telegraph traffic) or officials in the colonial service (who often possessed authority over the local telegraph office).[17]

Foreign ministries sometimes implemented policies that further deprived diplomats of the knowledge necessary to make independent decisions. The British Foreign Office initially attempted to telegraph only as much information as diplomats needed to accomplish their instructions. The Austro-Hungarian foreign ministry prohibited its missions in foreign countries from using the telegraph to converse with one another. They were allowed to use the telegraph only to communicate with their superiors in Vienna. This policy ensured that the Habsburg foreign minister had better access to information than his agents.[18]

Diplomatic Autonomy before the Telegraph

Slow communication and vast distances tended to increase the autonomy of diplomats during the era before the telegraph. The difficulty of promptly responding to unforeseen contingencies required sovereigns to invest great discretionary power in their envoys. In the Middle Ages, as one scholar has noted, "Treaties, truces, and other covenants made by medieval plenipotentiaries were complete and valid without subsequent ratification by the principal."[19] This sort of devolution of power continued into the nineteenth century. The 1840 Near East crisis, which pitted France against the other great powers of Europe over the future of the Ottoman Empire, demonstrates the enormous power accorded to distant envoys for dealing with emergencies. Although plenipotentiaries (rather than sovereigns) concluded the convention of 15 July 1840, in which all the powers of Europe except France attempted to pacify the Levant by compelling the ruler of Egypt to return recently conquered territory, it came into effect immediately, before the exchange of ratifications by their home governments. Because of the seriousness of the crisis and the distance that separated the contracting parties, the plenipotentiaries agreed, in the words of the treaty, "that the preliminary measures . . . will be immediately carried into execution without waiting for the exchange of ratifications."[20]

During the period before electric communication, governments wrote dispatches on the basis of old information and realized that their instructions would be even more dated when they finally reached their recipients. In the interim, many problems would have solved themselves. At times this made the entire exercise of writing instructions seem pointless, especially in a country that appeared to have few foreign interests. Thomas Jefferson, when serving as secretary of state under President Washington, is said to have observed that it was over two years since he had heard from the U.S. minister in Madrid. If another year went by with no news, he planned to write to him.[21]

Even when events overseas touched upon vital national interests, slow communication compelled governments to devolve power to diplomatic envoys. The mission of James Monroe to buy New Orleans from Napoleon is illustrative. As President Jefferson said, no instructions could be "squared to fit" the circumstances that Monroe would be facing. Likewise, during the crisis that preceded the Crimean War, Britain's foreign secretary gave the British ambassador in Constantinople a large say in one of the most serious decisions imaginable, whether or not to send his country's soldiers into combat:

> No later intelligence having since arrived, Her Majesty's Government are uninformed whether the Sultan . . . has actually declared war . . . [It] is difficult to send you precise instructions upon a state of things so critical, but with respect to which so much uncertainty still prevails.
>
> Her Majesty's Government are far from intending to impose upon your Excellency any responsibility that they can properly take upon themselves, but having explained the reasons which render precise instructions impossible, they think it advisable to leave to your Excellency and to Admiral Dundas, in concert with the French Ambassador and Admiral, to determine upon the best mode of giving effect to their views for the defence of the Ottoman dominions against direct aggression.

This devolution of authority is all the more striking since Britain's prime minister and foreign secretary expressed on numerous occasions their distrust of the British ambassador to Constantinople. If at all possible, they would rather have kept power in their own hands.[22]

Governments sometimes encouraged "disobedience" on the part of their overseas envoys. For example, the British government appears to have kept its colonial agents in the Americas poorly informed about military truces with Spain in the seventeenth and eighteenth centuries, presumably because the extra weeks of privateering were profitable.[23] Such Janus-faced policies proved useful when the demands of domestic public opinion differed from international exigencies. William Henry Seward, the U.S. secretary of state during the Civil War, went so far as to invite Charles Francis Adams, the American minister to Britain, selectively to ignore instructions from home: "Our instructions must always be based upon the understanding we have of facts at the time the despatches leave this department. On the other hand, the whole aspect of a case existing abroad is often changed without our knowledge, before instructions from this place are received, and sometimes, indeed, before they are written. In all cases you could hardly overdraw upon the confidence of the department in your wisdom and discretion."[24]

This statement accords with Seward's tendency to exploit the autonomy associated with slow communication for domestic political ends. Seward frequently wrote bombastic, Anglophobic dispatches, knowing that Adams was too sensible to deliver them verbatim to the British government. This allowed Seward to play a double game, writing bellicose warnings ostensibly aimed at Britain but actually designed to be published in America.[25] Seward thus managed to please the large anti-English component of the U.S. public without destroying Anglo-American relations; he knew that Adams would tactfully modify his instructions or ignore them completely. After Seward published some dispatches he had sent to Adams, Britain's foreign secretary wrote to his minister in Washington, "You are I believe aware that Adams never read to me any of those long, vainglorious + insolent despatches which Seward laid before Congress."[26] Such tactics worked. Once the British had become accustomed to Seward's bravado, it did not significantly damage relations between the two nations.

The independence that slow communication gave to diplomats brought with it the danger of disavowal. Consider David Erskine, who served as British minister to the United States from November 1806 until October 1809. This was a difficult time in Anglo-American relations: maritime disputes arising out of the Napoleonic Wars had produced great bitterness between the two countries. Erskine received his appointment from Charles Fox, a British foreign secretary and Whig politician noted for his friendly attitude toward the United States. Erskine's liberal attitudes and American wife made him a plausible choice to fulfill Fox's hope of improved Anglo-American relations. But Fox died in September 1806; Erskine lost his greatest supporter before he had even reached Washington. George Canning, who succeeded Fox as foreign secretary, showed considerably less interest in appeasing the United States. Tension between Erskine—a representative of the old policy—and Canning was inevitable.[27]

Canning did hope to improve Anglo-American relations, and he made limited concessions in this direction. For example, the British government revised the Orders in Council, the policy of commercial warfare against Napoleon, so that Americans were free to trade with Germany and the Baltic. But Canning wanted the Americans to assume an anti-French neutrality. He demanded that the United States open its ports to British vessels, end restrictions on commerce with Britain, agree not to trade with French-controlled areas of the globe, and accept British disciplinary actions against American ships caught trading with those areas. He wrote to Erskine on 23 January 1809 instructing him to resolve Anglo-American tensions along these lines. Erskine received these instructions in early April. Knowing that American political leaders would find the terms insulting, he decided to depart from Canning's precise directions. His negotiations with the Americans resulted in an exchange of notes on 18 and 19 April that met much of the spirit of his instructions. The agreement at which he arrived, however, did not meet Canning's demand for "a distinct and official Recognition of these Conditions from the American Government."[28]

With this accord, Erskine temporarily succeeded in attaining a significant improvement in Anglo-American relations. The agreement aroused exhilaration in the United States: the economic and psychological injuries inflicted by the Orders in Council would come to an end. One historian has written that "probably no other event from 1783 to 1815 was so generally celebrated." But Canning, after consulting with the cabinet, disavowed the agreement and wrote to inform Erskine of the impossibility "that you should continue in the Exercise of your Functions, either with Satisfaction to Yourself, or with Advantage to His Majesty's Service." The American public reacted bitterly to the repudiation of Erskine's agreement. Paul Hamilton, the U.S. secretary of the navy, said, "I am unable to express . . . the affliction and indignation I feel at the additional insult to the amicable disposition manifested by the United States towards Great Britain." Meanwhile, Erskine returned to Britain in disgrace, his geographic and psychological distance from his own government having led him to raise false hopes among the Americans. Such incidents led many diplomats to welcome faster forms of communication, which would allow them to seek permission and advice before taking an irrevocable action.[29]

Plausible Deniability

Governments have often found themselves needing to retreat gracefully in the face of opposition or to abandon unsuccessful policies while minimizing the loss of prestige. Before the widespread adoption of telegraphy, regimes possessed a simple expedient for dealing with such situations: they blamed their envoy for exceeding instructions.[30] In the world of intelligence and covert operations, such tactics have given rise to the phrase "plausible deniability," a tactic whereby principals disown the activities of their agents when the activities prove embarrassing or inconvenient. In such cases, an agent takes the blame in order to protect the boss and/or the organization.

The ability to stigmatize an overseas diplomat has sometimes provided cover for political leaders to take prudent but politically embarrassing decisions. James Polk's behavior as U.S. president during the Oregon crisis illustrates this approach to contentious decisions. In the 1844 presidential campaign, Polk implied that he would launch a war against Great Britain unless the United States received the entire Oregon territory up to the fifty-fourth parallel (hence the famous slogan "Fifty-four forty or fight"). Once in office, he showed some willingness to renege on his campaign promise and divide the Oregon territory with Britain at the forty-ninth parallel. When the negotiations encountered roadblocks, however, he and the newspaper that his administration controlled (Washington's *Daily Union*) again loudly demanded all of Oregon. In so doing, Polk gambled that Britain would retreat rather than go to war for acreage that Americans valued more highly than Englishmen.[31]

Louis McLane, the U.S. minister to Britain, overlooked Polk's bellicose statements and hewed to his instructions, which suggested a division at the forty-ninth parallel. McLane and Lord Aberdeen, Britain's foreign secretary, hammered out the basis for an agreement along these lines. Aberdeen sent it to Washington. He also included a British demand for continued navigation of the Columbia river, although this clearly violated Polk's stated demands. By the time the agreement reached Polk, on 6 June 1846, the administration had many reasons to seek a peaceful compromise of the Oregon issue. The American people no longer expressed the same enthusiasm over the prospect of war with Britain. Some of this mellowing resulted from the passage of time, which allowed an element of rationality to creep back into American public discourse. There were also reports that Britain, whose army and navy were vastly superior to those of the United States, had called Polk's bluff by ordering a military build-up.[32]

Meanwhile, U.S. forces were occupied by a war with Mexico that had begun a month earlier. In addition, American leaders knew that Robert Peel's government in Britain would soon collapse. This meant that at Britain's Foreign Office the cantankerous, anti-American Lord Palmerston—who would probably be more than a match for Polk—would replace the conciliatory, pro-American Aberdeen. Presumably for these reasons, Polk sent the Aberdeen-McLane agreement to the Senate for advice. But he protected himself politically by distancing himself from it and refusing to endorse it. McLane, annoyed by the president's lack of support, claimed that his own willingness to compromise with Aberdeen had been in line with the administration's policy. The result was a feud between Polk and McLane. Regardless of who was more correct, Polk's criticism of an agreement that his own agent had concluded gained credibility from the fact that slow communication made the control of distant envoys difficult.[33]

In theory, faster communication should have given principals greater responsibility for the conduct of their agents. When diplomats could be in instantaneous contact with superiors, it was more difficult for the home government to claim that a diplomat was following an independent policy. More difficult, but not impossible. April Glaspie, the U.S. ambassador to Iraq at the time of the 1990 Iraqi invasion of Kuwait, received considerable criticism for allegedly having departed from her instructions by encouraging Saddam Hussein to believe that he could get away with the invasion. This condemnation was apparently unjustified.[34] Yet it suggests that even in an age of rapid communication, diplomats continue to serve an important function by allowing governments and politicians to redirect blame or to change policies without losing face.

Technology and Diminished Autonomy

Machinery, such as the moving assembly line, the cash register, or the location tracker, has often facilitated employers' efforts to monitor and centralize the pace and control of work. In this way, machines have given management ways to heighten labor discipline, supplementing more traditional methods of control such as organizational or reward/punishment arrangements. In addition, such technology can be part of a process that transfers knowledge and control from skilled workers to management, thereby "proletarianizing" or "deskilling" workers and turning them into mere tools.[35]

Workers noticed management's efforts to tighten discipline. In many types of skilled employment, from glassmaking to diplomacy, they discussed how technology diminished their independence and debased human labor. According to an archetypal account from the glassmaking industry, "the cunning hand of the master craftsman" had once directed production, so that "the speed and rhythm of the work were set by the human organism, not by a machine." Then the development of glass-blowing technologies changed the job by lessening the need for human skill and making the workers appendages to the machines.[36] The introduction of the telegraph similarly reduced the autonomy of naval captains, who had once been highly independent but increasingly found that every port they entered possessed a telegraph line that connected them to their superiors. An American admiral reportedly declared: "The cable spoiled the old Asiatic Station. Before it was laid, one really was somebody out there, but afterwards one simply became a damned errand boy at the end of a telegraph wire." In the wake of the Suez crisis, a frustrated Royal Navy admiral exclaimed, "Nelson would never have won a single victory if there'd been a telex."[37]

In theory, the telegraph allowed closer supervision of diplomats. With the completion of the Atlantic cable in 1866, William Henry Seward, the U.S. secretary of state, sent a telegram whose high cost produced controversy. One reason it was so expensive was that Seward, with an eye to increasing his popularity among the American public, sought to create the impression that he managed events from afar. He later justified his long message by saying that he wanted his envoy to read it verbatim to Emperor Napoleon III. "For this reason," Seward noted, "no word was omitted upon any consideration of economy."[38] Although Seward implied that the cable destroyed the autonomy of the U.S. minister to France, reality was more complex—the minister decided not to deliver Seward's telegram.

Seward's verbosity was an exception: when foreign ministries used telegraphy to communicate with their diplomats, the desire to reduce costs generally promoted concise messages. The brevity of telegrams tended to make them stronger, less nuanced, and more authoritative than traditional dispatches. A British diplomat noted that instructions conveyed by telegraph were more likely "to be couched in peremptory terms than when prepared in the form of a Despatch."[39] While this was more a matter of form than substance, such telegrams may have induced timid envoys to display less independence.

Foreign policy institutions, when deciding how to deploy telegraphy, generally favored control over speed (see Chapter 4). They possessed legitimate fears about the confusion produced by speaking with two voices, or, as one policymaker put it, "the danger of attempting to negotiate at two ends of the line." In some circumstances the coming of telegraphy actually slowed the pace of diplomacy. For example, on 7 September 1870 Jules Favre, the French minister of foreign affairs, requested that the United States intervene diplomatically in the Franco-Prussian War to arrange a peace between the two countries. At the time, France was clearly losing the war, and this proposal would have increased pressure on Prussia to settle for a moderate peace agreement. Faced with Favre's request, the U.S. minister to France replied that he needed to refer the matter back to Washington, "particularly as I could advise with it on the subject by telegraph." An even more clear-cut case of telegraphy slowing the pace of diplomacy occurred at the beginning of this war when the Prussian ambassador to France asked the U.S. chargé d'affaires to assume protection over North Germans stranded in France. The chargé responded that while in the past he would immediately have acceded to the request, the existence of transatlantic telegraphy required him to contact Washington, despite the delay this would produce.[40]

The increase in centralized control had a profound effect on diplomats. With the advent of telegraphic dispatches, "the age of the 'great ambassadors,' who were perforce policy-makers in their own right, was over." Whereas chiefs of mission had once exercised great individual power as representatives of their sovereign, they now often felt themselves mere cogs in a bureaucratic machine, much like a British colonial agent who groused, "I might have been a great man, but for the telegraph." An article in the *New York Times* from 1900 describes the change:

> It is a fact that [the diplomat] has become less of a statesman and more of a correspondent, an exponent of his master's views, a go-between, an instrument. He is allowed little original action these days. Every move in his negotiations must be sanctioned by the Minister of Foreign Affairs at home, every objection or demand coming from the other side is referred to the same august autocrat. There are no Kaunitzes or Talleyrands in these times of the fast mail and the telegraph, when instructions more particular can be so easily asked, and supplementary orders so quickly transmitted.[41]

How did diplomats view the centralizing effects of telegraphy and the consequent reduction in their autonomy? Some, perhaps most, felt at times a degree of relief. Many diplomats, when facing confusing and unprecedented situations, felt burdened by responsibility, and leaned heavily upon their instructions.[42] For them, telegraphy provided welcome guidance. But there were also many complaints. A British envoy commented upon "this age of rapid communication, of what I would call the telegraphic demoralisation of those who formerly had to act for themselves and are now content to be at the end of the wire." Francis Bertie, who worked both in the British Foreign Office and as a diplomat, expressed a similar sentiment: "In Downing Street one can at least pull the wires whereas an Ambassador is only a d. . . .d marionette." Many diplomats felt that electric telegraphy deprived their activities of value.[43] Even the most powerful of ambassadorships seemed devalued. In 1913 President Woodrow Wilson offered Richard Olney command of the American embassy in London. Olney declined the offer, partly because his experience as secretary of state during the Cleveland administration had convinced him that "the English Ambassadorship is a show place—all the real work being done in Washington." Olney also noted: "An ambassador is nobody in these days. He sits at the end of a cable and does what he is told."[44]

Formerly, ambassadors could play major, perhaps even decisive, roles during international crises. According to one British diplomat, the telegraph changed all that:

> When crises of a certain magnitude occur, Foreign Ministers ignore their Ambassadors and correspond over their heads direct with one another . . . The system is hard on the Ambassadors, diminishes their stature and destroys the relics of such of their initiative as has been left them by the telegraphic tyranny under which they habitually live. It may, indeed, be necessary in order to gain great ends. But, if it is, it only goes to prove that in grave issues the diplomatic machinery to which we have so long been accustomed, is antiquated, and, given the modern facilities for quick communications between actual rulers, ineffective.

Two factors—crisis-induced stress and the desire to act quickly—made political leaders especially likely to circumvent their diplomats during international crises. These motivations to bypass diplomats had long existed. Telegraphy finally provided the means to do so: for the first time, governments could communicate directly with one another in something approaching real time.[45]

People outside the foreign service also noted the diminished role of diplomats. Finley Peter Dunne, an American humorist, offered homespun observations on how telegraph cables had reduced the privileged access of ambassadors to information: "In odher to keep up with what is goin' on th' Ambassadure has to get out arly an' buy th' mornin' papers. Th' first thing he knows about a war that's broke out between th' two counthries is whin he goes around to call on th' King an' a senthry jabs him with a baynit."[46]

Many citizens of the United States showed enthusiasm for the notion that telegraphy reduced the need for diplomats, perhaps rendering them altogether unnecessary. From the earliest days of the republic's history, Americans questioned the necessity of diplomatic representation. Thomas Jefferson saw diplomats as generally expendable. John Adams wondered whether Americans ought to "recall their ministers and send no more." During the presidency of George Washington, some senators sought to prevent the establishment of an American system of permanent diplomatic representation abroad.[47] In line with this tradition, politicians used the onset of electric telegraphy to advocate the abolition of the diplomatic service. Samuel Cox, a congressman from New York, declared: "Telegraph, steam, with their prompt communications; newspaper enterprise ever in advance of diplomatic dispatch; these and other elements of progress have rendered ministers abroad trifling, expensive, and useless for every purpose of national comity, interest, or glory." In 1858, following the successful laying of a (short-lived) transatlantic cable, a minister in Boston extolled it as the "greatest discovery of this age." He declared that traditional diplomats were more expensive than the cable and frequently promoted mistrust between nations, "darkening counsel by words." He believed nations would be better off when they communicated directly with one another rather than relying upon unfaithful diplomats.[48]

Such sentiments also found expression in Britain. Cox's critique of the State Department claimed inspiration from parliamentary debates about the "nonsense, extravagance, and inutility" of a diplomatic service "in this age of steam and telegraphs." In 1861 a parliamentary committee charged with reforming the British diplomatic service had felt obliged to consider whether traditional diplomats had lost their usefulness: "A question has . . . been raised . . . as to whether the rapid means of communication now established by railway and telegraph does not render the transaction of diplomatic business more easy; and that, in fact, the work might be done by the Foreign Office maintaining a mere clerk or agent, to transact the business under its own direction, without having to pay high salaries to diplomatic agents." The Foreign Office managed to blunt these attacks, however, and such criticism became less common over the course of the nineteenth century as mid-Victorian parsimony gave way to late-Victorian imperialism.[49]

Technology Is Not Destiny

While telegraphy facilitated centralized control of diplomats, it did not make this control inevitable. In the short run (which, in the conservative world of diplomacy, could last for decades), institutional culture played a more important role than technology in determining the autonomy of a country's diplomats. During the years before the First World War, Germany and France provide an illuminating contrast in this regard.

Otto von Bismarck shaped the institutional culture of Germany's foreign ministry, the Auswärtiges Amt, by insisting upon absolute obedience from his subordinates during his long tenure as chancellor. The

British ambassador to Berlin observed, "Any attempt at independence or initiative he stamps out at once." Bismarck did not tolerate outright disobedience in any sphere of life. He squelched the politically unsatisfactory marriage plans of his eldest son, Herbert, and permanently embittered the young man. He likewise became infuriated with the tendency of his favorite dog, Sultan, to stray. In a rage, Bismarck beat and mortally wounded the mastiff. Filled with remorse, he was disconsolate for days thereafter. As for German diplomats, Bismarck once remarked, "My ambassadors must wheel about at command like non-commissioned officers, without knowing why."[50] Although Bismarck disdained bureaucracy, even prizing a genial nonchalance among his subordinates, he demanded conformity in matters of policy. He acted decisively against diplomats who refused to follow orders. Count Georg Herbert zu Münster lost his post as ambassador to Britain because he refused to comply with Bismarck's wishes. The chancellor's war against Harry von Arnim, the German ambassador to France, became so fierce that Arnim was imprisoned before fleeing into exile. The chancellor eloquently articulated the reasoning behind his demand for diplomatic obedience: "Observation at close range has its advantages and also its disadvantages, and where it is a question of forming a judgement on a question . . . which encroaches on the field of general policy the best foundations for a sound estimate are as a rule to be sought at the centre where responsibility lies." Even after Bismarck's forced departure from office in 1890, this philosophy continued to influence the German diplomatic corps, which subsequently exercised little independence. Telegraphy was a useful tool for Bismarck and his successors as they wielded control over German diplomats.[51]

The conduct of French diplomats before the First World War differed greatly from that of their German counterparts. Paul Cambon, France's ambassador to Britain and a champion of the Anglo-French entente, exercised particular influence over French foreign policy. Devoted to his country and its culture—one historian states that he believed the ability to speak English "showed a cultural deficiency"—Cambon possessed utter confidence in his ability to interpret and defend his country's national interests. He had little faith in the ability of the citizenry, or the politicians they elected, to guide French policy. He sometimes burned his instructions when he found them objectionable. A great believer in the autonomous judgment of capable diplomats, he thought that France's ambassador to Berlin might have prevented the Franco-Prussian War had he delayed the implementation of his instructions.[52] Cambon derived considerable power from his informal leadership of a faction that included France's powerful ambassadors to Rome, Washington, and Berlin (the French ambassador in Berlin was his younger brother, Jules). Working together, these ambassadorial mandarins effectively challenged the decisions of bureaucrats at the foreign ministry and imposed their views upon the highest French politicians. The autonomy of French diplomats resulted from their self-confidence and ability, a lack of faith in their government (the Third Republic was chronically unstable), and, perhaps most important, the institutionally weak position of the French foreign minister. The centralizing effects often attributed to telegraphy were not conspicuous within the French foreign ministry.

Resistance

Workers frequently dislike oversight, especially when their opinions or interests conflict with those of their bosses. The history of labor suggests as much: workers have often sought to circumvent, or even sabotage, technologies that deskilled them or reduced their autonomy on the job.[53] Military history presents similar examples, as when naval captains resisted the introduction of radio. In occupations such as diplomacy, in which the accomplishments of employees are difficult to assess, workers and managers are especially prone to friction.[54] One might expect diplomats to have obstructed use of the telegraph, especially when they disagreed with the policies supported by their superiors. The challenge, as in other cases of secret resistance by workers, lies in uncovering incidents of insubordination that diplomats sought to conceal in order to avoid reprimands.

Telegraphy tended to concentrate power and promote bureaucratic hierarchy. Independently minded diplomats were not without recourse, however. The high cost, insecurity, and technical shortcomings of the telegraph gave them serviceable excuses for not using it at inconvenient moments. The actions of Lord Stratford de Redcliffe (Stratford Canning), Maurice Paléologue, and Baron Wilhelm von Schoen demonstrate the possibility of resistance by diplomats to the telegraph's centralizing tendencies.

Stratford Canning (1786–1880), known after 1852 as Lord Stratford de Redcliffe, displayed a genius for exploiting the characteristics of the telegraph, such as its high cost, to preserve his autonomy. He is an almost ideal archetype of the self-reliant diplomat who seemed to reign in the age before electric telegraphy. Yet he lived long enough to comment upon the beginnings of telegraphic diplomacy. A choleric, Russophobic ambassador who represented Britain in Constantinople during much of the early and mid-nineteenth century, Stratford has been termed "the last diplomat to hold the issue of war or peace in his hands." Indeed, some have blamed him for the outbreak of the Crimean War. Many of Stratford's contemporaries, including Lord Aberdeen (the prime minister), the Earl of Clarendon (the foreign minister), and Queen Victoria, accused him of encouraging the Turks to assume a belligerent posture that eventually dragged Britain and France into a war with Russia.[55]

While most recent scholarship has tended to exonerate Stratford from the more extreme accusations of responsibility for the war, he was far from an obedient ambassador. His enormous ego, boundless self-assurance, fiery temper, and cantankerous personality continually vexed his superiors. Aberdeen pondered how to exercise control over him: "I think it will be necessary to be very careful in preparing instructions for Lord Stratford." But Stratford had a way of artfully misinterpreting his instructions, and Aberdeen complained, "I thought that we should have been able to conquer Stratford, but I begin to fear that the reverse will be the case and that he will succeed in defeating us all." Stratford thus presents a textbook example of the sort of diplomat that foreign ministries longed to control through better means of communication.[56]

Stratford's career lasted just long enough to extend into the era of telegraphy. He received the first telegram sent from London to the British embassy in Constantinople, apparently on 31 August 1855. Fortunately for historians, his testimony to a parliamentary committee in 1861 elucidated his opinions about the telegraph's influence upon his independence. On the whole, his views challenge standard notions of technological determinism. For example, Stratford declared that telegraphic communications were "subject . . . by their very nature, to the risk of conveying erroneous information, or premature instructions." Furthermore, telegrams were "liable to frequent mistakes in the transmission" and were sometimes so concise (because of cost) that diplomats were forced to rely upon their own judgment.[57] Stratford, drawing upon his experiences during the 1850s, asserted that the problems of unreliability and concision hindered telegraphic diplomacy and necessitated a "superior judgment" (such as his own) to decide whether or not a given message from the Foreign Office should be disregarded. Such remarks provide evidence that Stratford's tendency toward insubordination would have continued, telegraph or no telegraph.[58] And yet the telegraph would have allowed officials in London to learn of and respond to Stratford's independent courses of action more quickly.

A number of British officials who looked at matters from London rather than from Constantinople agreed with Stratford that the price of telegrams undermined efforts to micromanage foreign policy from afar. In 1861 Lord John Russell averred that the brevity inspired by the high cost of telegraphy gave diplomats more autonomy than they would have had if reliant upon long, detailed, written instructions. Sir John Tilley, chief clerk of the Foreign Office during the First World War, observed, "It is certainly the case that an Ambassador may be a good deal at sea as to the reason for his instructions and the spirit in which they are issued, and yet may not feel that he can delay action till he has obtained clearer explanation."[59] Compared with written dispatches, telegrams carried a small amount of information very quickly. This fact required foreign ministers to write succinct instructions that made their intentions clear; when they were incapable of doing so, diplomats often gained more leeway than one might have expected them to possess in an age of electric telecommunications.

Diplomats of a later generation tended to be less forthright in their resistance than Stratford. The mystery surrounding Maurice Paléologue's (1859–1944) activities during the crisis that preceded the First World War should feature in any discussion of disobedient diplomats. Paléologue was, in the words of one historian, "a self-dramatizing and unstable romantic" who had previously been the political director at the French foreign ministry on the Quai d'Orsay. In early 1914 he was appointed ambassador to St. Petersburg, despite French President Raymond Poincaré's fear that his Russophilia made him unsuitable for the post. Paul Cambon, France's ambassador to Britain, supported the shift in part because he believed Paléologue would be able to do less harm to France's foreign policy in the new assignment. At an earlier posting, an Austrian military attaché had noted Paléologue's "fantastic imagination [which he] permits to run away with him when he interprets insignificant military or political events, and, for those who do not know him well, he is therefore dangerous as a source of information."[60]

Like many French diplomats of his day, Paléologue felt that he was accountable only to "France." He also believed in a "great man" theory of history, and had no qualms about disobeying instructions. Most dangerously, Paléologue combined a deep pessimism about German intentions—he perceived a Franco-German war as inevitable—with a conviction that the military balance currently favored the Franco-Russian alliance. The logic of his beliefs led to the wisdom of a war with Germany. Furthermore, Paléologue had some knowledge of Germany's military plans and knew that, in the event of war, German armies would quickly descend upon his country in an effort to defeat France before Russia could mobilize. This information, that Germany intended to crush the allies separately, created a powerful incentive for Paléologue to hasten Russian mobilization against Germany during a crisis.[61]

Paléologue's behavior during the "July crisis" that preceded the First World War has met with severe criticism on two grounds. First, historians have accused him of disobeying his instructions by encouraging Russian belligerence at a time when his superiors sought to encourage Russian restraint. Second, historians have reproached him for keeping Paris ignorant of Russian decisions during the crisis.[62] French policymakers, poorly informed by Paléologue, were unable to lobby against Russian actions that heightened the conflict, such as first partial and then general military mobilization. When authorities in Paris finally acted, their initiatives were too late. For example, on 31 July 1914 René Viviani, France's prime minister, sent Paléologue a telegram that ended: "As I have already informed you, I have no doubt that the Imperial [Russian] government, in the greater interest of peace, will avoid, for its part, anything which could begin the crisis." But by this time Russian mobilization had already begun and war was inevitable. Because of their ignorance of events, French political leaders had no opportunity to influence the decisions made by their alliance partner, decisions that would have far-reaching effects on the future of their own country. These criticisms of Paléologue, if they are accurate, imply that even in the era of the telegraph—when power and knowledge became increasingly centralized—distant diplomats sometimes had enough autonomy to subvert the foreign policies of their countries, perhaps even bringing on a global war.[63]

The criticism about Paléologue's role in keeping the government in Paris ignorant of events in Russia has provoked a rejoinder that sheds light on the question of diplomatic autonomy in the age of the telegraph. Those historians who defend Paléologue suggest that rather than deliberately keeping his government uninformed, he refrained from transmitting some messages by telegraph and sent others by an indirect and extremely slow Scandinavian route, because of fears about security.[64] Paléologue may have worried that messages encoded in the French code and transmitted along the normal line, which passed through Germany, would be vulnerable to German signals intelligence, which perhaps knew the French code. In addition, defenders of Paléologue argue that the Russian government shared his fears about the insecurity of French diplomatic messages and therefore deliberately withheld information from him in order to delay German knowledge about Russian military mobilization. This could explain why Paléologue sent one of his telegrams to Paris via the Russian foreign ministry, which transmitted it in a Russian code.[65]

A number of factors cast doubt upon the contention that fear of German codebreakers reading French telegrams explains Paléologue's failure to inform his government about the Russian moves toward war. First, evidence for this contention has been derived from only two sources: Paléologue himself and Nicolas Basily of the Russian Foreign Ministry.[66]

Second, given the supposed insecurity of Paléologue's code, why did he not ask the Russian government to use its embassy in Paris to send news of its military mobilization to the French government? All known accounts that allege the insecurity of the French embassy code also assert that Russian encryption methods provided satisfactory security against German codebreakers. The use of Russia's code would have allowed the Russian government to inform France of its mobilization measures, which the Franco-Russian treaty obliged it to do. The fact that this was not done suggests that other factors were at work that had little to do with the supposed weakness of the French code.[67]

Third, the fear of an insecure code cannot explain Paléologue's failure to report important information *before* the evening of 29 July 1914, at which time he reportedly learned about the vulnerability of French communications. To cite one of many examples, Paléologue waited a day before telegraphing the Quai d'Orsay to inform it of the Russian Council of Ministers' decision on 24 July to call for a partial mobilization in the event of an Austro-Hungarian attack on Serbia.[68]

Fourth, if Paléologue was concerned about the security of his dispatches, he should have directed his suspicions toward his ally Russia, which intercepted, decoded, and read French diplomatic telegrams. Germany, in contrast, did not intercept diplomatic telegrams before the First World War.[69] In any case, Paléologue's poor reporting resulted in his own government, rather than Germany, being slow to learn of Russian mobilization. The German government found out about it almost immediately. The Russian plan was impossible to conceal since it involved putting up posters around St. Petersburg announcing mobilization. Paléologue's negligence produced an absurd situation: France's prime minister, Viviani, learned of Russia's mobilization from the German ambassador to France (who mentioned it to him while inquiring into French policy in anticipation of the coming Russo-German War).[70]

Fifth, Paléologue's failures to keep his government satisfactorily informed continued *after* Russia's general mobilization had become public knowledge and any rationale for further secrecy about it had disappeared.[71]

Together, these arguments suggest that Paléologue sought to keep the Quai d'Orsay unaware of Russian foreign policy in order to increase the chances of war. Yet caution is in order. Merely reconstructing Paléologue's behavior during the July crisis is quite difficult, given his duplicity and mercurial personality. Divining the motivations behind his actions is an order of magnitude more complicated. Should we believe that he ineffectually sought to fool (nonexistent) German codebreakers about Russian mobilization, or rather that he successfully kept his own government ill informed until war became inevitable? I subscribe to the latter view. But in either case, the potential insecurity of telegrams gave Paléologue a plausible excuse, regardless of whether or not it was truthful, for resisting the centralizing effects of the telegraph. As a result, he managed to shape the information reaching his superiors, thus concentrating power in his own hands during the crisis.

The experiences of Baron Wilhelm von Schoen (1851–1933), Germany's ambassador to France before the First World War, illustrate the way the telegraph's tendency to garble messages could undermine central control of diplomats. Born a bourgeois, Schoen strove to display aristocratic manners to accompany the title he had acquired in his mid-thirties. Within the German foreign service, Schoen had a reputation as a "society man" who used his courtliness and geniality to avoid unpleasantness.[72] For this sort of diplomat, communication problems could furnish a handy mechanism for eluding bothersome directives while also evading a disagreeable confrontation. The frequency of scrambled telegrams made such excuses plausible. Everyone knew that garbled messages sometimes brought a welcome respite from unwanted orders.[73] In some cases, critics had accused independently minded diplomats of sabotaging the policy of their government by slyly garbling telegrams that they received.[74]

On 3 August 1914 Schoen presented a German declaration of war to the French prime minister, in anticipation of Germany's invasion of France. The German government, in a telegram, had instructed Schoen to justify this extreme measure by citing hostile acts committed by French aviators over Germany and alleging that French ground troops had crossed the German frontier. Schoen, however, discussed only the former grievance, claiming that the part of the telegram discussing French border incursions was jumbled and unintelligible.[75]

The French air strikes that Schoen cited to justify Germany's attack turned out to be imaginary. This declaration of war, based on false grievances, made Germany's invasion of France—legally difficult to justify in any case—even more ridiculous. Schoen later complained: "An adverse fate . . . obliged me to confine myself to statements which gave the French Government abundant ground for asserting that we had trumped up excuses to justify our attack. Not only the Press, but also Ministers, and particularly M. Poincaré, have delighted in making use of this effective means of proving that we made an unprovoked attack." Schoen also wrote, "The telegram was so mutilated that, in spite of every effort, only fragments of it could be deciphered." He added, "As I knew from other sources that we felt bound to declare war in consequence of a French air attack on Nuremberg, I had to make up my mind to fall back on the little that could be clearly understood from the telegram, to justify the declaration of war."[76]

Schoen's avowal that he possessed alternative sources of information about intentions in Berlin is tantalizing and possibly refers to intelligence from the Habsburg ambassador to France. In any case, Schoen must have had great confidence in this mysterious source to justify declaring war on France despite the unintelligibility of his instructions. Perhaps, as he asserts, he lacked time to wire his government asking for a repetition of the mutilated part, although he had asked his government to repeat a garbled telegram earlier that day; but one would at least have expected him to inform his superiors in Germany that the lack of clear instructions, due to a garbled telegram, was forcing him to declare war on his own initiative. Furthermore, the German government had sent the telegram by two routes, and a second, ungarbled copy reached Schoen at 5:00, presumably leaving enough time for a message of such importance to be deciphered before Schoen's meeting with the French prime minister at about 6:30 P.M. Moreover, Schoen telegraphed the German government at 8:30 that evening, incorrectly stating that he had received and executed his instructions. He neglected to mention the garbled telegram that, he would later allege, had prevented him from delivering the war declaration in its original form.[77]

Did Schoen have an incentive not to deliver the war declaration as written? He had, one historian notes, just received a protest from the French about an incident in the village of Joncherey, in which German troops had killed a French corporal on French soil, "and now he was to be compelled, without making the slightest explanation, and under instructions from his Government, to declare that not a single German detachment had crossed the frontier! How could he reply to the remarks which M. Viviani would be sure to make upon this point? He thus had a personal interest in omitting the passage concerning the violation of the territorial frontier." Schoen later became involved in disputes with French scholars about his conduct during the July crisis; he clearly cared about his reputation. On 3 August 1914 he may have calculated that it was better to base the declaration of war on German complaints that might be true (although they turned out not to be) than on protests that he knew were hypocritical for reasons that were probably not known in Berlin (since the German political leadership was unlikely to have heard so soon about the Joncherey incident).[78] While this argument cannot be proven with certainty, there is suggestive evidence that Schoen falsely claimed that the telegram conveying the German declaration of war had been garbled in order to spare himself the obloquy of presenting a dishonorable war declaration. In any case, Schoen's behavior, like that of Stratford and Paléologue, demonstrates a tactic that diplomats could use to evade the closer oversight that accompanied the adoption of telegraphic diplomacy.

Diplomats have frequently been targets of public wrath. If, as seems to be the case, xenophobia is a natural human tendency, then individuals who habitually associate with foreigners will inevitably be subjected to charges of disloyalty. In addition, some critics have felt exasperated with the arrogant and constricted attitudes of the privileged social strata from which diplomats tended to be drawn. Many have also resented the elitist trappings of diplomacy: genteel manners, elaborate apparel, and a life of professional leisure (filled with receptions, dinners, and balls). Telegraphy and other communication technologies created a focal point for these feelings. Critics asked what purpose diplomats served when international communication had become so easy. The decline in the independence of diplomats exacerbated the public conception of them as decadent and idle. Given these criticisms, how is it that the institution of foreign-based diplomats endured?

Custom and inertia are part of the explanation for the survival of foreign legations in an age of electric communication. But there have also been practical reasons for the maintenance of overseas missions. First, diplomats contribute valuable perspective: the view from abroad differs from the view at home. Recent developments in communication (such as e-mail and teleconferencing) have expedited this role, increasingly allowing diplomats not only to provide input for, but also to participate directly in, the policymaking process at home.[79]

Second, diplomats help governments make sense of the messages they exchange. Telegrams, to a greater extent than written dispatches, separated their explicit content from the physical surroundings in which they had been created: the recipient of a telegram did not see the original paper submitted to the telegraph clerk. Furthermore, telegrams tended to be extremely concise because of concerns about expense and speed. The medium of telegraphy produced highly abstract messages devoid of the vocal inflection, facial expression, or gesturing likely to occur in face-to-face interaction. As a result of this lost context, telegraphy eliminated much useful information from messages.[80] Diplomats helped restore this information, which would have been lost had governments merely interacted via the telegraph.

For example, diplomats could give their governments information that would make incoming messages comprehensible. Acting as skilled listeners, they facilitated governments' efforts to assess one another's activities. Diplomatic envoys had opportunities to interact socially with the leaders of the countries in which they were stationed, often resulting in private, semi-inebriated conversations. Such access gave them a good vantage point for making character judgments—which are often based on nonverbal behavior—about political leaders.[81] This sort of information was crucial for evaluating missives from those governments in which information was tightly controlled and a small number of people wielded considerable power.

Diplomats also played a vital role in the conveyance of dispatches from their own governments. They "performed" messages, and thereby rendered them more understandable to foreign leaders. For example, the absence of context made it difficult to tell whether a telegram was authentic. Austria-Hungary's declaration of war against Serbia in 1914 demonstrated this problem. Because the Habsburg government had withdrawn its minister from Belgrade, it had no means of personally delivering its war declaration to Serbian political leaders. The Dual Empire decided to send the declaration to the Serbian foreign ministry by telegraph (via Romania, since Austria-Hungary had cut the lines between Vienna and Belgrade). The Serbian prime minister, surprised by the message and the unusual procedures of its delivery, initially believed it a hoax.[82]

Third, organizational expertise developed in other sectors of society—particularly in business—provides a rationale for decentralization that can be used to justify the existence of diplomats stationed abroad. The correct balance to strike between centralization and decentralization has been called the "central problem of management."[83] The subject of business organization does not lend itself to easy generalization, but corporations have repeatedly found that complete centralization of decisionmaking authority is impossible, and, were it possible, would pose many problems. Although the deskilling of workers diminishes the bargaining power of labor, thereby allowing for wage cuts and a more obedient work force, in most areas of business some degree of worker initiative and delegation of authority makes the organization better able to adapt to change and alleviates pressures on higher managers.[84] This was even more true in the field of diplomacy, where complete centralization of decisionmaking posed difficulties because of the varied, unpredictable, and sensitive nature of many problems.

Although no immutable rules exist concerning the proper degree of centralization, some general observations on the subject have merit. Centralized control often facilitates coordination, efficiencies, and economies; a wider distribution of power and responsibility allows for initiative, a sense of accountability, development of personnel, decisions close to the facts, and flexibility. The effort to reconcile these goals—order and autonomy—poses challenges for every organization. In general, organizational theory suggests that complex decisions be moved (up, down, or sideways) within an organizational hierarchy until they reach those individuals possessing the greatest access to relevant information. Furthermore, other matters being equal, it is often efficient to push decisions downward in order to relieve upper management of unnecessary demands.[85]

These principles were developed through trial and error. During the nineteenth century most government bureaucracies, including foreign ministries, still attempted to concentrate decisionmaking authority at the top, and they adopted new technologies, such as telegraphy, with this goal in mind. A historian of Britain's Colonial Office aptly articulated the dangers of centralization that the telegraph promoted: "Questions great and small, which formerly were decided on the spot, were now thrown upon the Home Government, not with the full information of a despatch, but with the elliptical curtness of cipher telegrams."[86] Centralization, pursued as an end in itself, tended to place a heavy burden of decision on the center, regardless of whether the pertinent information was available there. Diplomats based in foreign countries generally had better access to local information than did their colleagues at home, and this privileged knowledge allowed them to act with greater efficiency when dealing with purely local affairs. The elimination of diplomats would have impoverished the decisionmaking abilities of the organizations for which they worked.

Fourth, foreign-based envoys served purposes that, while far from glorious, were eminently practical: as built-in delays and as scapegoats. Neither of these functions depended on diplomats possessing the power to make independent decisions. When leaders felt that events were proceeding too quickly, diplomats could be relied upon to produce deliberation and helpful inefficiency. Even in those unusual cases where overseas envoys did possess the authority to commit their government, they had a stronger negotiating position if they kept this secret. For this reason, Henry Kissinger said that he "resented bitterly" the full powers that he received from U.S. President Richard Nixon to negotiate a ceasefire at the time of the 1973 Mideast War. Kissinger's annoyance was not with the accretion of authority, but rather with Nixon's decision to inform the Soviets of it. Once Kissinger could no longer claim that he was referring matters back to the president, he lost one of his best excuses for postponing decisions until a more advantageous moment.[87]

Diplomats also make good scapegoats when governments wish to repudiate unsuccessful policies. By taking responsibility for errors that they did not commit, diplomats furnish a relatively graceful mechanism for governments to change course or correct mistakes. A British official described these functions:

> We have spoken of the diplomatist as a fender, and it is in fact part of his duties to be one. By functioning as an intermediary he provides the makers of policy with time to think. International business is usually most intricate and full of pitfalls; every move is liable to have far-reaching consequences. Direct negotiation between principals is therefore apt to be dangerous by reason of its finality. It gives no time for reflexion, and there is little possibility of retrieving mistakes once they have been made. The diplomatist as intermediary can always take the blame, and not infrequently does. When any back-pedalling is necessary in foreign policy he can always say to the Minister for Foreign Affairs: "I think I may

> have failed to put your point clearly enough to my government, so give me time to explain it to them again." He does so (or at any rate appears to do so); and with the most gratifying results. The rub has been taken by a fender; the principals have not lost face.[88]

For such reasons, telegraphy, and the telecommunications technologies that have succeeded it, while substantially changing the lives of diplomats, have not destroyed their profession.

Speed II

The Trent Affair

3

In the autumn of 1861, public opinion in the northern part of the United States had reached a low ebb. The initial engagements of the Civil War had demonstrated the emptiness of the North's "on to Richmond" bravado. Even the most sanguine observers were beginning to comprehend the stupendous effort and suffering that a military defeat of the South would entail. Foreign exaltation at Southern successes added insult to injury. British mockery of the Union's martial spirit—as when the London *Times* gleefully referred to "the swift-footed veterans of Bull Run"—created particular irritation in the North because of the long history of Anglo-American animosity, the accessibility of British newspapers, and the basis in reality of British derision. European hostility toward the Northern cause boded ill. The American secretary of state, William Henry Seward, worried that "the pressure of interests and ambitions in [London and Paris] have made it doubtful whether we can escape the yet deeper and darker abyss of foreign war." Reports about Confederate diplomatic missions to Europe exacerbated these fears of foreign intervention. After so much bad news, Unionists desperately sought a victory of some sort.[1]

For their part, British observers watched the Civil War with mingled horror, fascination, and hope of gain. Anglo-American economic and social integration meant that events in the United States (such as the cessation of Southern cotton exports and the North's large arms purchases) had repercussions in England. Southern secession provided unexpected opportunities for Britain to establish a balance of power in North America, safeguard the security of Britain's American colonies, and rein in the territorial and commercial expansionism of the United States. Many Englishmen believed that intervention on the side of the South was justified from a moral perspective. In their view, Lincoln's initial refusal to abolish slavery robbed the struggle of its moral significance and made it seem one of conquest and needless slaughter. Moreover, the North's demand for high tariffs threatened British interests and violated the Victorian ideal of free trade.[2]

Lord Lyons was Britain's minister to the United States in the autumn of 1861. The lack of a working Atlantic telegraph cable (the one laid a few years earlier had failed) made his position important and influential, particularly during that period of diplomatic crisis.[3] He had received early diplomatic training as a child when he and his three siblings pretended they were an embassy staff, writing diplomatic dispatches on behalf of "His Excellency," the ambassador, impersonated by the family dog. Adult diplomacy was more frustrating for Lyons. Shy, conscientious, fastidious, and a gourmand, the Englishman disliked life in Washington. The city, he observed, was "hot and dull," while the dinners on the diplomatic circuit were "interminable in length, miserable in quality and not often amusing in the way of society." Respite had seemed possible a few months earlier, following the South's victory at Bull Run. Despite his antipathy toward slavery and the Southern cause, Lyons cheerfully speculated that Confederate troops might be able to occupy Washington and force the diplomatic corps to relocate to another city, a prospect that he imagined "might be rather agreeable." Lyons would later find a silver lining in the death of the queen's husband, Prince Albert, which required a period of mourning on the part of the British legation. Abstention from society dinners, the envoy remarked in a letter to his sister, "has been no sacrifice."[4]

Yet Lyons found Secretary of State Seward even more ominous than Washington's social circuit. The reserved British diplomat and the excitable American politician seemed ill suited to work together. Seward cultivated a reputation for being bellicose, aggressive, and a loose cannon. Moreover, he spent much of his time devising schemes to expand the commerce and territory of the United States; he had recently expressed an interest in adding Britain's North American possessions (present-day Canada) to the American empire.[5] As a senator, Seward had promoted the construction of telegraph lines and railways.[6] He believed these tech-

Lord Lyons, British minister to the United States from 1859 until 1864.

Library of Congress, Prints and Photographs Division, LC-USZ62-14978

nologies played a crucial role in integrating and extending American power. As he declared, "Empire moves far more rapidly in modern than it did in ancient times."[7]

Most foreign observers believed that Lincoln deferred to Seward in the making of foreign policy, especially since the president had remarkably little experience in that realm. Lyons wrote gloomy reports informing his superiors in London about the man who would direct the conduct of American diplomacy: "I cannot help fearing that he will be a dangerous Foreign Minister. His view of the relations between the United States and Great Britain has always been that they are a good material to make political capital of. He thinks . . . that they may be safely played with without any risk of bringing on a war." During the course of 1861, the need for Britain to take a resolute stance against Seward's bluster was a recurring theme in Lyons's dispatches. In May Lyons reported that "Mr. Seward is so arrogant and so reckless towards

William Henry Seward, secretary of state under Abraham Lincoln and Andrew Johnson.

Library of Congress, Prints and Photographs Division, LC-USZ62-21907

Foreign Powers" that taking "a firm stand" against him was the only way to keep him "within bounds." Several months later Lyons suggested that improved American conduct "has been mainly produced by our preparations for defence and by the quiet firmness with which we have maintained [our rights]."[8]

Lyons's assessment of Seward was not entirely off the mark, as can be seen from a famous memorandum Seward wrote on 1 April 1861. In this note to Lincoln, the secretary of state seemed to advocate making demands upon Britain and Russia while launching a war against Spain and France. Perhaps Seward, under heavy emotional pressure resulting from the cabinet decisions that produced the first fighting at Fort Sumter, genuinely believed that he could divert America's violent passions away from civil conflict into a foreign war. Seward may also have realized the usefulness of being seen as a madman: European countries would be less likely to aid the Confederacy if they believed that the North would respond with irrational ferocity. In any case, British statesmen heeded Lyons's counsel and anticipated the worst from the Lincoln administration.[9]

The scene was set for another crisis in Anglo-American relations. The irascible Charles Wilkes, captain of the American navy's steam frigate U.S.S. *San Jacinto,* provided it. Wilkes knew that James Mason and John Slidell, the Confederate commissioners to London and Paris, had eluded the Union's blockade of Charleston, reached Cuba, and taken passage for St. Thomas (where they planned to board a ship to Southampton) on the British mail steamer *Trent.* In response, the *San Jacinto* slipped away from Cuba a few days before the departure of the *Trent* and waited for its quarry. At noon on 8 November 1861, three hundred miles east of Havana, the *San Jacinto* halted the English ship with two shots across the bow. Captain Wilkes ordered a boarding party to arrest Mason and Slidell and their secretaries, seize their possessions and any Confederate diplomatic dispatches, and make a prize of the *Trent.* However, the boarding party did not find the diplomatic dispatches and disobeyed Wilkes's instructions by not commandeering the *Trent.* Wilkes acquiesced in this decision, apparently swayed by the arguments that seizing the mail steamer would inconvenience many passengers and irritate the British and that the departure of the crewmen necessary to convey the *Trent* to the United States would weaken the *San Jacinto* in future naval engagements. Thus the *Trent* continued on its way while the *San Jacinto* sailed toward Boston, reaching Fort Warren on the evening of 23 November to deposit the Confederate prisoners.[10]

Wilkes arrived in Boston to a hero's welcome. Massachusetts Governor John Andrew gave the captain a public dinner. Journalists and eminent jurists vied with one another to praise and justify his interpretation of international law. Adulatory crowds met Wilkes as he made a triumphant procession southward toward Washington. When the U.S. House of Representatives met on 2 December, it promptly awarded him a gold medal. After a year of repeated disasters, Unionists finally had an excuse to celebrate. Wall Street provided one of the few sources of pessimism as stocks that were sensitive to Anglo-American conflict declined.[11]

Lincoln and his cabinet responded with pleasure to the news of Mason's and Slidell's capture. The lone dissenting voice was Montgomery Blair, the postmaster general, who warned that the Confederate commissioners would have to be released. Lincoln and Seward wisely avoided committing themselves to either side of the controversy until they learned the content of the British reply. On 15 December, however, Seward attended a ball at the Portuguese legation at which, perhaps under the influence of brandy, he declared that if Britain forced war upon the United States, "We will wrap the whole world in flames!" Meanwhile, Lord Lyons, the voice of Britain in the United States, maintained a much-noted reserve as he waited for instructions from his government.[12]

On 27 November news of the *Trent* incident reached London. George C. Lewis, Britain's war secretary, predicted that the crisis would produce "inevitable war." Lord Palmerston, the prime minister, reportedly began a meeting with some of the members of his cabinet on 28 November by throwing his hat on the table and declaring, "I don't know whether you are going to stand this, but I'll be damned if I do!" Across the English Channel, Emperor Napoleon III, keen on good relations with Britain and enthusiastic about the possibilities of French intervention in America, strongly supported Britain during the crisis.[13]

On 30 November Lord Russell, Britain's foreign secretary, composed an ultimatum that demanded an apology and the return of the Confederate diplomats. The rest of the cabinet edited the dispatch and passed it on to Queen Victoria and her husband, Prince Albert. By this time the British government had delayed the departure of a large mail steamer so that it would be able to carry the final version of the dispatch. The Prince Consort, mortally ill, found the response drafted by the British government too bellicose. He softened the wording without changing the essential demands. The cabinet accepted Albert's modifications and sent out the ultimatum with an accompanying dispatch instructing Lyons to give the United States seven days to accede to the demands after he had delivered them. While Lyons could "be rather easy about the apology," the release of the Confederate diplomats was essential. Unless they were freed, Lyons was to close the British legation in Washington, breaking off diplomatic relations and conveying an implicit threat of war. At the last moment Russell added a private note that gave Lyons additional flexibility regarding the time limit: Seward should be told about the ultimatum at the first meeting, but the seven-day period would not begin until a second meeting, at which Lyons would formally present the demands to Seward. Russell also suggested that Lyons delay presenting the British demands for "a day or two" to allow time for France's diplomatic support for the British position to reach Washington. The mail steamer *Europa* departed with the dispatches on 1 December.[14]

Palmerston's cabinet accompanied these political decisions with military precautions. It prohibited the export of saltpeter (the main ingredient in gunpowder) and weapons to the United States. More dramatically, it sent troops to defend British North America, although some believed these colonies were indefensible against an attack from the United States.[15] The Admiralty ordered a concentration of the fleet and prepared to blockade the North. While the British government did not seek war, the *Trent* affair provided it with a litmus test of the North's intentions. The Earl of Clarendon, a former foreign secretary with ties to Palmerston's cabinet, wrote, "If the Americans are determined upon [war], and if, as I believe, they have been looking for an opportunity to quarrel with us, the sooner it comes, the better, as we are not likely to have a better case [than the *Trent* affair] to go to war about, nor shall we ever be better, or they worse, prepared for war." Lyons's dispatches from Washington had preconditioned British policymakers to expect a crisis of some sort, and his warnings about Seward led the British foreign secretary to write, "The best thing would be if Seward could be turned out, and a rational man put in his place."[16]

Lyons received his instructions from the foreign secretary late on 18 December. He visited Seward at the State Department the next day and summarized the ultimatum. Seward asked whether the British government had stipulated a time limit for a response from the United States. Lyons told him of the seven-day deadline. Seward queried whether he might have a copy of the dispatch "unofficially and informally," since this would save a great deal of time and much depended on the exact wording. Lyons replied that if he officially gave Seward the dispatch, the seven-day countdown would begin, but Seward responded that this was unnecessary since only he and the president would know that Lyons had given him a copy. Lyons noted, "I was very glad to let him have it on these terms. It will give time for the Packet (which is indeed already due) to arrive with M. Thouvenel's Despatch to M. Mercier [the instructions for the French Minister in Washington], and in the meantime give Mr. Seward who is now on the peace side of the Cabinet time to work with the President before the affair comes before the Cabinet itself." Seward went to the British legation almost immediately after receiving the dispatch and noted his pleasure at finding it "courteous and friendly, and not dictatorial and menacing."[17]

Seward and Lyons met again on 21 December, at which time Seward asked for an extension of two days for the commencement of the countdown. Lyons wrote Russell that he agreed to this request because of the irregular state of transatlantic relations: "No time was practically lost by my consenting to the delay . . . for whether the seven days expired on Saturday next or Monday next, I should have been equally unable to announce the result to you sooner than by the packet which will sail from New York on Wednesday, the 1st January." On Monday, 23 December, Lyons officially presented the ultimatum.[18]

Meeting on Christmas Day, the cabinet was unable to make a decision about the Confederate diplomats. While the meeting was in process, Henri Mercier, the French minister to Washington, rushed to the State Department with the information that he had finally received the long-awaited instructions from his government, which the French foreign minister had sent on 3 December. This note, showing French solidarity with Britain, was introduced into the meeting, where it ended the argument that the North would be able to play off Britain and France against each other. Yet Lincoln did not wish to comply fully with the British demands. The cabinet members decided to discuss the matter again the next day. After four hours of sometimes heated discussion, that second meeting ended with the decision to surrender Mason and Slidell. Seward wrote a lengthy and legally specious reply to Lyons that managed to be both conciliatory enough to appease the British—by allowing for the return of the prisoners—and assertive enough to placate the American public.[19] Lyons obtained this message on 27 December and sent it to his government. The Confederate diplomats were released on 1 January 1862. They received a cool reception in England.

The *Trent* affair was one of the last major international crises in which the capitals of the participating countries lacked a telegraphic connection with one another. Although extensive land telegraph systems served economically advanced areas such as Western Europe or the United States, engineers had not yet developed the technology to telegraphically link continents separated by great oceans. Compared with later crises, the *Trent* affair seemed to occur in slow motion. In 1861 a diplomat had to wait about a month to receive an answer to a dispatch he had sent across the Atlantic.[20] Although the U.S. Navy apprehended Mason and Slidell on 8 November 1861, the British complaint about the seizure did not reach Lyons in Washington until 18 December. This incident provides a valuable source for examining the effects of telecommunications upon crisis diplomacy because some of the actors themselves explicitly speculated about the impact that a transatlantic cable might have had upon the crisis.

In 1858 the British and American publics had celebrated the completion of an Atlantic telegraph cable. This cable never worked well, and completely stopped functioning after less than three months.[21] The effort to reestablish telegraphic communication between Europe and America did not succeed until 1866, after the end of the U.S. Civil War. In the interim, many contemporary observers wondered how such a device might have affected the outcome of the *Trent* affair. Some asserted that a functioning transatlantic cable would have allowed for the quick resolution of the crisis, nipping it in the bud before it could develop into a war scare. For example, Cyrus Field, one of the principal supporters of a transatlantic cable, wrote to Seward in early 1862 that "a few short messages between the two governments and all would have been satisfactorily explained." In contrast, others argued that an Atlantic cable would have made war more likely by depriving the participants of the time necessary to settle the dispute. Lord Lyons himself frequently asserted that the existence of such a cable during the *Trent* affair would have made war inevitable.[22] Since then, historians have sometimes mentioned this issue, generally making brief assertions that either Field or Lyons was correct.[23]

Let us first consider Cyrus Field's view: that during a diplomatic crisis, faster is better because it prevents a misunderstanding from worsening. The proposition that the slow speed of transatlantic communication during the *Trent* affair made war more likely can be broken down into two arguments: that a slower pace of international relations reduced room for diplomatic maneuver; and that slow communication caused participants to develop an excessively negative view of the other side. When measured against the historical evidence, both arguments seem incorrect in this case.

The first argument asserts that fast communication expedites diplomatic negotiation. The *New York Herald* seemed to make this point when it declared that the entire war scare would have been prevented had the British government immediately known the contents of Seward's dispatch of 30 November.[24] According to this view, while waiting for feedback, political leaders and publics become trapped by the adoption of policies that reduce negotiating options. George C. Lewis, the British war secretary, articulated this contention during the crisis: "I am very much inclined to think that if a submarine telegraph from England to New York, or Portland, or even Halifax were now in working order, the chances of war would be greatly diminished. If war occurs, it will be owing to the Americans committing themselves irrevocally [*sic*], before they know the serious view of the transaction which is taken in England + France."[25]

In addition, as crises continue, participants and observers alike may become emotionally engaged and increasingly resistant to making the compromises necessary for successful negotiations. Perhaps for this reason, John Bright, a prominent political figure in Britain, warned an American contemporary in early December 1861 that "nations *drift* into wars, . . . often thro' the want of a resolute hand at some moment early in the quarrel." Indeed, one historian argues that Northern opinion did become more adamantly self-righteous during the wait for the British response to the capture of the Confederate diplomats.[26]

Nonetheless, there is good reason to doubt the claim that slow communication hindered the North's ability to offer diplomatic concessions during the *Trent* affair. Lyons, who, as Britain's diplomatic representative in Washington, was much better situated to judge the effects on the North's diplomacy than was Lewis, believed the delay gave Americans time to weigh the dangers of war with Britain. Seward, too, must have believed that slowing down the crisis facilitated his diplomacy; otherwise he would not have asked Lyons to extend the time limit. In addition, scholars who have analyzed public reaction to the *Trent* affair have tended to argue that Northern opinion moderated and became less obdurate as the crisis continued and that British opinion also shifted toward a more conciliatory posture.[27]

Psychological analysis of foreign policy decisionmaking suggests a second argument (again supporting Field's view) by which transatlantic telegraphy might have facilitated the resolution of the *Trent* affair. Statesmen tend to exaggerate the animosity of other countries. Such misperceptions become self-fulfilling prophecies: mistrustful statesmen behave in a hostile manner toward states they regard as antagonistic, and the other states respond with hostility, thus confirming the original misperception.[28] In this manner, the absence of information about the intentions of others can produce a vicious spiral that results in unnecessary international conflict and even war. By keeping statesmen better informed, faster communication may reduce the likelihood of conflict arising from ignorance.

The *Trent* affair does not support this theory. In neither the United States nor Britain did the view of the other country become more negative during the delays caused by the telegraph. Rather, as shown by the studies of public opinion mentioned earlier, each country began with the belief that the other was fairly hostile and, during the period of waiting, developed views of the other that, while still somewhat negative, were more nuanced and hopeful.

Lord Lyons's view, that the slow speed of transatlantic communication facilitated a peaceful resolution of the *Trent* crisis, can be divided into three arguments about the slower pace of diplomacy before the telegraph: that it led to more thoughtful foreign policy decisions, that it reduced the likelihood that an enraged public would push policymakers toward war, and that it protected political leaders from being overwhelmed by excessive information. The *Trent* affair supports the first two of these arguments and provides some evidence for the third.

In support of the proposition that telegraphy posed dangers during crises, one might begin with the argument that faster communications can lead to hasty and defective decisionmaking. In the midst of a crisis, many people feel an urge to do something, regardless of whether immediate action is warranted. Lyons wrote that during the *Trent* controversy he had resisted the temptation "to do something," which "always besets one when one is anxious about a matter." The *Trent* affair provides evidence that the first reaction to dramatic political events is not necessarily the wisest. One historian has noted that "passion, expediency, and poor judgment seemed to have infected almost everyone." Charles Francis Adams Jr., the son of the U.S. minister to Britain, was living in Boston in November 1861. Writing fifty years later, he recalled the public reaction to the capture of the Confederate diplomats:

> I do not remember in the whole course of the [past half-century] any occurrence in which the American people were so completely

> swept off their feet, for the moment losing possession of their senses, as during the weeks which immediately followed the seizure of Mason and Slidell . . . Morbidly excited and intensely sensitive, the country was in a thoroughly unreasoning and altogether unreasonable condition . . . for it needed only the occurrence of some accident to lead to a pronounced explosion of what can only be described as Anglophobia.[29]

One should also remember that Montgomery Blair was the lone member of Lincoln's cabinet initially opposed to the seizure and that the others only slowly accepted the necessity of freeing Mason and Slidell. Similarly, the British response moderated during the days before Russell dispatched it to Washington. Both sides would have shown themselves much less conciliatory if they had acted upon their initial impulses. Such evidence suggests that the leading participants benefited from having time to reflect upon the situation.

The second argument that bolsters the view that faster communications pose hazards is the tendency of an impassioned public to dominate the policymaking process. This proposition is in some respects the opposite of the argument that compromise becomes more difficult the longer a crisis lasts. Rather, the emotional reaction to dramatic events makes compromise more difficult in the immediate aftermath of a crisis. Previously cited evidence of the increasingly peaceable trends of opinion in Britain and the United States over the course of the *Trent* affair offers support for this view.

The third argument suggests that a sudden acceleration of communications can impair the work of diplomats by producing information overload. New technologies often require organizations to adapt themselves to changing capabilities and requirements. There is generally a lag between the challenges posed by a technology and successful institutional responses. The shift from ship-carried to electrically transmitted messages increased both the number and the constancy of demands upon diplomats during international crises. Foreign ministries moved toward a "continual-response mode" whereby governments felt a need to respond quickly to events around the world. The result could be physical and mental exhaustion.[30]

Even without transatlantic telegraphy, the U.S. Civil War, then in its first months, placed great strains on the U.S. State Department and the British legation in Washington at the time of the *Trent* affair. Seward's biographer has described him as "often overworked." This is hardly surprising since U.S. secretaries of state were still responsible for personally reading and replying to all dispatches of importance. The British legation in Washington also confronted a vastly increased workload because of the war. Junior members of the legation frequently became unwell and even Lyons's health eventually collapsed, probably because of excessive labor and the unhealthful climate.[31] A functioning transatlantic cable, especially one that was relatively recent (such as the cable of 1858, if it had worked properly), might have increased the pressures upon and impaired the functioning of the diplomatic institutions of both countries during the *Trent* crisis, with deleterious effects on Anglo-American relations.

The *Trent* affair provides an instance when the use of the telegraph in diplomacy, by eliminating periods of delay, would have created more problems than it solved. During intervals of waiting, initial reactions could be questioned, participants could compose themselves, and plans could be reconsidered, all with generally beneficial results. There is little evidence to support the argument that accelerating the pace of the *Trent* affair would have facilitated diplomacy. Of course, slow communication made substantial demands upon diplomats. A successful resolution of the dispute required both governments to avoid committing themselves as they waited for the situation to develop.

Consider the performance of the two diplomatic representatives, Charles Francis Adams and Lord Lyons. The threat of war placed each under close scrutiny, while the absence of an Atlantic cable gave both a great deal of responsibility. Although neither could singlehandedly settle the controversy—the matter being too important for a diplomat to take decisive action without instructions—either of them could have greatly aggravated the conflict by a careless word or a lack of poise. Most of all, both managed to keep a low profile and avoided becoming swept up in the passion of dramatic events.[32] Britain's foreign secretary later praised Lyons for his "silence, forbearance and friendly discretion."

What was the philosophical underpinning of Lyons's conduct? One of his grandnieces recalled hearing him assert that the most valuable axiom in diplomacy was "Never do anything to-day that can be put off till to-morrow."[33] Many of Lyons's peers acted in accordance with this precept. What would happen when diplomats espousing such views found themselves in a world where information traveled at the speed of electricity?

Speed and Diplomacy

4

Observers have frequently exaggerated the speed of telegraphy. They have even claimed that it completely obliterated the notion of distance. But even a more judicious appraisal grants great significance to this technology. Telegraph wires and cables sent messages at the speed of electricity, and thereby separated communication from transportation.[1] For the first time, complex messages sent over long distances could travel faster than people could proceed by ship, horse, or train. The separation of message and messenger created new possibilities for active control of distant physical processes in real time.[2] Telegraphic speed affected the efforts of political leaders and diplomatic envoys to cope with international crises. By its impact on both the role of public opinion and the quality of decisionmaking, telegraphy tended to make the management of diplomatic crises more difficult, and arguably increased the likelihood of war. But before addressing these issues, let us first consider the extent to which the telegraph actually accelerated international relations.

Faster Diplomacy?

The electric telegraph made possible a rapidity of communication that fascinated contemporaries. In 1846 a newspaper in Rochester, New York, declared, "The actual realization . . . that instantaneous personal conversation can be held between persons hundreds of miles apart, can only be fully attained by witnessing the wonderful fact itself." According to the parlance of the day, "telegraphic" became a synonym for quick. Observers marveled at the supposed annihilation of time and space, which represented an "extraordinary victory over the powers of nature."[3] An American periodical proclaimed, "There are no longer any far off lands . . . Steam and electricity have well-nigh blotted out distance." London's *Daily Telegraph* quipped, "Time itself is telegraphed out of existence." The speed of the world, observed Henry Adams, "when measured by electrical standards as in telegraphy, approached infinity, and had annihilated both space and time." A French scholar wrote, "Diplomatic agents are today, like everyone, suspended by telegraph wires; space and time no longer exist."[4]

Such statements lacked technical accuracy. For example, telegrams sent over long submarine cables moved quickly but were far from instantaneous because the surrounding water conducted electricity, slowing and blurring the signals. This problem required that telegraphers operate at a speed of, on average, about thirty words per minute.[5] Not until the 1920s and the laying of "loaded" cables (a means to keep the signal sharp), was this problem largely overcome. But fervent statements about the annihilation of distance did capture the extraordinary acceleration that electric telegraphy made possible in long-distance communication. At the same time, the existence of a capability does not guarantee that it will be fully exploited. A number of nontechnological factors limited telegraphy's impact upon the speed of diplomatic intercourse. The three most important stumbling blocks were the desire of foreign ministries to control their distant agents, the intricate procedures involved in sending and receiving diplomatic telegrams, and the traditions, habits, and whims of the human beings who staffed the world's foreign policy institutions.

The first factor, the desire of foreign ministries for greater control, in combination with the adoption of telegraphy, sometimes slowed the course of diplomacy. Before electric communication, the need to respond quickly to distant events often forced governments to delegate authority to representatives. Faster communication enabled foreign ministries to more closely supervise their subordinates stationed abroad. As a result, the telegraph reduced the individual initiative of diplomats and frequently retarded the course of international relations. In some cases after the establishment of transatlantic telegraphy, American diplomats, faced with problems they would previously have resolved themselves without delay, believed they had to seek guidance from the State Department because they could now do so without waiting weeks or months for a response. As U.S. minister to France in 1870, Elihu Washburne found himself "asking for instructions, though not doubting what would be the prompt action of our government." In an earlier era, he would have decided such matters on his own. The telegraph created the possibility or even the obligation to refer matters to the foreign ministry rather than acting independently. This resulted in delays.[6]

The second factor, the intricate procedures involved in diplomatic telegraphy, meant that it did not provide instant transmission of information, contrary to what was widely assumed. The myth of instantaneous communication has sometimes fooled historians, scrambling the chronology of fast-paced events in which timing was important, such as the crises that preceded the Franco-Prussian War and the First World War.[7] The drafting and revision of telegrams took time. Diplomats tended to choose every word in a telegram with great care because of the need for concision (due to cost concerns) and the dangers of ambiguity. The Earl of Clarendon, who served Britain as a diplomat and foreign secretary, commented, "You must remember that a telegram is a condensed despatch; I have had more trouble sometimes in writing a telegram than I have in writing a despatch." Moreover, through the period of the First World War, foreign ministry clerks manually coded and decoded important telegrams, a laborious process that could significantly delay messages. In the case of a message of particular importance, a foreign ministry official might even cable a warning to a distant diplomat: "You will receive a very important message as soon as it can be put into cipher."[8] The seriousness of cipher-related holdups led foreign ministries to send uncoded messages when speed was crucial and secrecy unimportant.[9]

After encoding, messages might sit for some time before they were taken to the telegraph office. Once there, they waited in a queue until after the transmission of preceding telegrams. Relay stations, which received and retransmitted messages to boost the signal, further slowed messages covering large distances. In 1900 a cablegram from London to Australia took about six hours. Moreover, telegraph companies often repeated encoded telegrams, since such messages were incomprehensible to telegraphers and therefore prone to garbling. They might also recapitulate transmissions when they had reason to believe an error had occurred. The manager of a telegraph company warned the German embassy in Washington, "Due allowance should be made for the time required for transmission and repetition." Extra delays resulted when efforts to avoid garbling failed. A U.S. diplomat in Costa Rica complained that decoding one telegram occupied him and his wife for nearly a day and another telegram took him and the American consul "the better part of two days." After a message was received and decoded, it would often be processed by the bureaucracy and therefore not reach its target immediately. In 1998, when the British Foreign Office announced that it would stop using telegraphy, newspapers noted that a telegram not marked "urgent" typically required twenty-four hours to reach the intended minister.[10]

Of all the diplomatic telegrams that have ever been delivered late, the most famous is probably the final message from the Japanese government to U.S. Secretary of State Cordell Hull, breaking off negotiations before Japan's commencement of hostilities on 7 December 1941. Because the communication reached Hull after the attack on Pearl Harbor, it became a symbol of perfidy that strengthened American resolve and stimulated the war effort. Members of the Japanese elite, arrested after the war and tried for war crimes, defended themselves against the charge of a surprise attack on the United States by citing the incompetence of the Japanese embassy in Washington, which, they claimed, had unnecessarily delayed the telegram's presentation.[11] Japan's diplomats in Washington thus received blame for besmirching their nation's honor.

A closer examination of this event demonstrates many of the reasons why diplomatic telegrams were sometimes delayed. In Japan, speculation about the delay has centered on rumors of heavy drinking the night of 6 December by embassy officials—who threw a farewell dinner for one of their colleagues—and a late arrival at work the next day. But other factors also deserve mention. The Japanese foreign ministry had not alerted the embassy by designating the message as urgent. By 1941 Japanese diplomats used decoding machines to accelerate the work of code clerks. But the Washington embassy, as a security precaution, had destroyed two of its three machines on the orders of the foreign ministry, greatly slowing its ability to decipher messages. Another security measure prevented the professional American typist from handling the message, resulting in a great loss of efficiency. Exacerbating the situation, the telegram had arrived with many garbled and missing words, which retarded the work of the clerks. And the schedule was tight. The last section of the fourteen-part telegram arrived at 7:30 a.m. (Washington time). The Japanese diplomats learned at 10:00 a.m. that they were supposed to deliver the telegram to the State Department at 1:00 that afternoon. In the end, the frantic Japanese diplomats did not finish preparing the memorandum until 1:50 p.m. and did not reach the State Department until 2:05, about forty minutes after the Pearl Harbor attack had begun.[12] In the age of the telegraph, communication delays were always possible and sometimes probable. The leaders in Tokyo designed a constricted and, under the circumstances, unrealistic timetable. They, much more than the diplomats in Washington, bear ultimate responsibility for the late missive.

The human element provided a third factor that sometimes prevented diplomats from receiving information in a timely manner. Regardless of what circumstances seemed to dictate, individual human beings ultimately decided whether or not to exploit the rapidity of telegraphic communication. In 1881 U.S. Secretary of State James G. Blaine sent William H. Trescot, a State Department official, on a special mission to bring an end to the War of the Pacific. In this conflict Chile had decisively defeated the forces of Bolivia and Peru, but the peace talks dragged on inconclusively. Blaine believed that American pressure from Trescot might induce Chile to offer peace terms that would be acceptable to its vanquished foes. Blaine, however, lost his position after the assassination of President James Garfield. The new secretary of state, who disagreed with the policy of anti-Chilean pressure sent revised instructions to Trescot by steamer. He also published Trescot's old instructions and publicly repudiated them. The Chilean minister to the United States telegraphed news of these events to Santiago (via Paris). As a result, the Chilean government received word of the modification of Trescot's instructions before Trescot did. Trescot learned of the change during a discussion with the Chilean foreign secretary, who took glee in Trescot's ignorance and discomfiture. Trescot was reduced to pleading, "If I am to receive my instructions through you let me know them in full." The U.S. government, by publicizing secret information and choosing not to employ the telegraph, had embarrassed and discredited its own envoy.[13] This incident demonstrates that even in an age of accelerated international crises, human beings ultimately decided (wisely or otherwise) when its use was appropriate. They were not mere tools of the technology.

Although the telegraph did not inevitably or uniformly quicken the tempo of diplomacy, two factors—the increasing power of public opinion and the pressures of international crises—ensured that it did tend to accelerate international relations. The growing influence of public opinion on foreign policy compelled governments to keep themselves well informed and capable of rapid action. Such pressures encouraged the use of the telegraph. For example, in 1851 the French ambassador in London expressed enthusiasm about the new Anglo-French cable, arguing that French diplomats and policymakers "needed to be informed [about important events] at least as soon as the public." Two decades later, in 1870, the Foreign Office chided Britain's ambassador to France for slow reporting of disturbances in Paris described by British newspapers: "Will you let me suggest to you that it would have been convenient if you had sent us a telegram yesterday on the subject, stating how the matter stood. It would have been awkward if a question had been asked in Parliament, founded on what appeared in the paper, + no reply could have been given on your authority."[14]

The dynamics of international crises also invited the greater use of the telegraph and, in consequence, the acceleration of diplomacy. Crises tended to place psychological stresses upon decisionmakers that encouraged them to act with haste. Furthermore, at least in theory, the telegraph removed the primary obstacle to rapid diplomacy during crises. International crises tend to occur at the pace at which political leaders of the countries involved can communicate with one another. Political leaders, when facing issues of war and peace, know they will be held accountable for failed policies. Not surprisingly, they usually respond to these circumstances by concentrating power in their own hands; they are often unwilling to grant diplomats in distant locations the power to make major decisions. In 1853 Britain's foreign secretary, the Earl of Clarendon, expounded upon this point after his subordinate, the temporary head of the British embassy in Constantinople, summoned a British fleet to the Dardanelles without instructions from home: "You may suppose that we are not a little anxious to know what induced Rose to send for the Fleet, as we think that questions of peace and war had better be decided by the Government at home than by a Chargé d'Affaires in the far East."[15]

Before the telegraph there tended to be a great distinction between crisis and noncrisis diplomacy. During normal, noncrisis periods in an age of relatively slow communication, foreign ministries typically gave their distant agents broad guidelines regarding their duties but allowed them considerable leeway in interpreting this guidance. After all, no instructions could anticipate all contingencies, and most problems would have solved themselves before new instructions could be sent. But during crises foreign ministries typically withdrew this freedom, and governments communicated more directly with one another rather than allowing their envoys to represent them. Without the devolution of control to on-the-spot intermediaries, crises could become long and drawn out as governments waited for answers to diplomatic initiatives carried by ships or messengers. The existing media of long-distance communication tended to impart their particular speed and rhythm to severe international disputes (see Chapters 1 and 3).

In the age of the telegraph, diplomatic crises could play out with startling rapidity because of the new potential for speedy communication between distant capitals. One sees the influence of electric telecommunications in the brevity of the diplomatic crises that preceded the Franco-Prussian War and the First World War. The pressure to act quickly during crises—which emanated from both personal psychology and journalistic/public opinion—ensured that statesmen would seek to exploit these new capabilities of the telegraph, sometimes unwisely.

Public Opinion

The movement of public opinion during foreign policy crises of the nineteenth and early twentieth centuries is impossible to measure with any precision. Reliable polling did not exist until the 1930s. Nonetheless, it is possible to consider data or theories about public opinion from later periods and extrapolate them back onto the historical record, and to examine the evidence that does exist on contemporary public opinion.

Scholars have given some thought to how public opinion varies over the course of diplomatic crises. The research, as one might expect, indicates that publics become more excitable during crises. This response is generally temporary, however. People frequently return to their former views once the excitement wears off. An apparent example of this phenomenon is the "rally 'round the flag" effect, whereby U.S. presidents sometimes become more popular with the public during a major international crisis or event. Regardless of what causes the "rally" effect and how often it occurs—both of which are matters of debate among scholars—there is little dispute that it tends to be a short-term phenomenon. Although presidents can gain considerable popularity during these crises, the effect tends to be "of very short duration, averaging a little over three months." One study found that twenty-six of forty examples of public "rallying" (65 percent) returned to pre-event levels within ten weeks, while 40 percent lasted five weeks or less.[16]

Although there is much we do not yet know about the evolution of public opinion during international crises, the evidence indicates that the passing of several weeks or months can significantly affect public opinion, perhaps because of the often temporary nature of the emotional shock and changed patterns of information flows that accompany crises. The implication is that a telecommunications technology such as the telegraph, which tended to reduce the duration of international crises from months to weeks or even days, confronted decisionmakers with a more agitated public. It would have been possible for political leaders to rally public support in such circumstances, but only if they seemed to have a plan for managing the crisis. The public cannot easily perceive quiet diplomacy, whereas military action tends to project strength and decisiveness, at least in the short run. It therefore seems plausible that shorter crises may have increased incentives for political leaders to take hasty and belligerent actions during crises.

A second tactic for analyzing telegraphy's influence upon public opinion is to consider historical incidents. A brief examination of this subject suggests that delays due to slow communication tended to reduce public belligerence, particularly when this belligerence resulted mainly from a specific event. In other words, the passage of time acted as a pacifier, if not always a conciliator.[17] As the *Trent* affair suggests, the delays in diplomacy produced by ship-carried dispatches provided time for tempers to cool and peacemakers to go about their work. A similar phenomenon played out in 1807, when a British attack on an American naval ship, the

Chesapeake, infuriated the American public and seemed to bring the two countries to the brink of war. Many historians, with this incident in mind, have noted that the United States had stronger grounds for fighting Britain in 1807 than it did in 1812, when it actually did go to war.[18] Indeed, U.S. public enthusiasm for war was much greater in 1807 than it was five years later. Yet despite white-hot initial rage over the assault on the *Chesapeake,* the temper of the American public cooled while President Thomas Jefferson methodically collected information about what had happened, presented the U.S. demands to Britain, and waited for the British reply. The American public's clamor for war had largely receded within a month of the attack. The inevitable delays attendant upon transatlantic communications provided a publicly acceptable explanation for the president's delayed response. Meanwhile, public war fever dissipated and the likelihood of a war that had seemed imminent slipped away, despite the fact that Jefferson's demands to British authorities went largely unmet.[19]

The anger aroused among the U.S. public during the First World War by Germany's torpedoing of the *Lusitania* may seem to offer a counterexample to the argument that slowing the pace of events allowed an enraged public to calm down after a dramatic event, thereby decreasing the likelihood of war. On 7 May 1915 a German submarine sank the *Lusitania,* a British liner, killing 1,198 people, including 124 U.S. citizens. The American public responded with shock, horror, and anger. This reaction forced President Woodrow Wilson to demand that Germany disavow the sinking, pay reparations, and renounce surprise attacks on unarmed merchant vessels. In contrast, the German public expressed jubilation at the sinking; such sentiment constrained the ability of moderates within the German government to implement policies that would soothe American sensibilities. By the end of the year the dispute had still not been resolved, and the U.S. secretary of state hinted that war might result if it continued to fester. Although American talk of time urgency can be seen as a negotiating tactic designed to pressure German leaders into acquiescing to U.S. demands, there seems little doubt that war was then possible, despite the passage of half a year since the initial provocation. The drawn out nature of the crisis did not reduce the risk of war.[20]

What this analysis misses, however, is the fact that much of the continuing tension between the United States and Germany over the *Lusitania* occurred within a context of additional sinkings that repeatedly reignited American anger.[21] These fresh injuries diminished the conciliatory effects of the passage of time. Nonetheless, delaying tactics did contribute to the resolution of the immediate crisis. On 17 May 1915, slightly more than a week after the disaster, Count Johann von Bernstorff, the German ambassador to the United States, warned his government of the importance of avoiding "another event like the present one." He instructed his superiors that Americans, when they are "deeply moved and excited . . . can be compared only to an hysterical woman, to whom talking is of no avail. The only hope is to gain time while the attack passes over . . . I only hope that we will survive it without war." Years later, in his memoirs, Bernstorff recalled his stratagem for dealing with the crisis by seeking to "gain time, during which the waves of indignation might die down."[22]

Britain's interruption of Germany's underwater telegraph cables impaired Bernstorff's ability to contact his government, and, in this instance, greatly facilitated his diplomatic efforts to decelerate the pace of the crisis.[23] He temporarily placated the U.S. government, which complained about Germany's slow response to the crisis, by arguing that no more could be expected given his difficulties in communicating with Berlin.[24] Meanwhile, his delaying tactics created a crucial interval during which his allies in the German government could produce pro-American concessions. As one historian has written, Germany's chancellor, who sought peace with the United States, "had time both for maneuver in imperial councils and for manipulation among domestic political groups."[25] The resulting compromise between the two governments preserved peace until Germany abandoned this agreement and adopted unrestricted submarine warfare in 1917. Although Bernstorff's efforts ultimately failed, he helped postpone war between Germany and the United States for nearly two years. His achievement owed much to his understanding of the emotional surges that impelled U.S. public opinion.

The French scholar Albert Sorel, writing in the late nineteenth century, asserted that telegraphy deprived publics and governments of time for reflection. During crises, this resulted in impassioned publics demanding precipitate action. To illustrate his view, he compared the international crisis of 1840, when France considered but rejected a foolish war, with that of 1870, when France fought and lost such a war. In both crises, issues of prestige, fears of domestic political instability, and expansionist impulses impelled the French government toward a policy of military conquest at German expense. But the first crisis occurred before diplomatic telegraphy and persisted over many months, whereas the second is famously associated with telegrams (especially one from the resort at Ems) and ran its course within a couple of weeks. Sorel described the "popular delirium" in favor of war that gripped Paris during the crisis that preceded the Franco-Prussian War of 1870. And he noted that in 1840 the French public, suffering from similar illusions and lust for war, considered entering a conflict that would probably have led to national disaster. Yet in 1840, Sorel contended, the absence of electric communication allowed time to think, and in the interim "the nation recognized the gravity of the crisis."[26] The decline in French bellicosity over the course of the 1840 crisis lends credence to Sorel's argument, as do the claims of French statesmen in 1870 that they felt pressured by an "overexcited" public to take an aggressive stance against Prussia.[27] As Sorel asserts, the telegraph posed dangers to peaceful relations between states. Faced with a volatile public, political leaders felt tempted to pursue aggressive policies that would increase their popularity in the short run.[28]

A related claim is that electric telegraphy began what would later be called "the CNN effect": the notion that dramatic reporting can shape public opinion and thereby drive foreign policy. During the 1990s it appeared to some that television news coverage in such disparate and distant locations as Somalia and Bosnia not only shaped public opinion in the United States but also set the U.S. foreign policy agenda. But claims that the media drive foreign policy are not new. For example, historians have found it difficult to eradicate the myth that the battle between the Hearst and Pulitzer papers produced the 1898 Spanish-American War.[29]

The assertion that telegraphy increased the influence of the media and public opinion over foreign policy possesses some merit. Telegraphy encouraged governments to adopt activist foreign policies because they now had a greater ability to affect distant events as they occurred. This new capability, especially when dramatized by the newspapers, inspired public interest in foreign affairs. Moreover, faster communication created a sense of urgency and an expectation that governments should act decisively. In 1889 an editorial in Britain's *Spectator* warned of the dangers of a rapid dissemination of information to the public via the telegraph: "The result is a universal haste and confusion of judgment, a disposition to decide too quickly, an impatience if hurried action is not taken before the statesmen or others responsible have had time to think. It is as if all men had to study all questions under the excitement of anger or fear or pity, or that conscious sense of being 'bustled' which . . . makes true reflection most difficult or impossible." The public excitement and the sense of urgency imparted by telegrams encouraged interventionist foreign policies.[30]

Nevertheless, there is a danger of overstating the changes brought about by the telegraph. Even during times of relatively slow communication, public opinion still pushed governments toward foreign intervention. The main basis of the CNN effect is the vividness of television, not its speed in reporting information, which is hardly new. Yet a well-written newspaper article could also be vivid and arouse the public, regardless of how it was conveyed.[31] This is illustrated by an incident that took place before the Crimean War: on 30 November 1853 a Russian fleet chased a Turkish squadron into Sinope, a Turkish port on the Black Sea. The Russians smashed the Turkish ships and destroyed much of the port city as well. The news of this action did not reach London until 12 December, but when it did, it outraged the British public and provided an impetus for the British fleet to begin anti-Russian patrols in the Black Sea.[32] It thus appears that the "CNN effect" preceded not only CNN but also other electric media, including telegraphy.

Pressure on Diplomats

Telegraphic speed had consequences for the decisionmaking abilities and well-being of diplomats, foreign ministry officials, and statesmen. It also affected the functioning of foreign policy institutions. In 1864 Charles Francis Adams, the U.S. minister to Britain, feeling overburdened by the intense diplomacy associated with the American Civil War, speculated on how the proposed Atlantic Cable would affect his job:

> Although entirely friendly to [the cable] scheme, I must confess I am not very anxious it should be carried out immediately. It is a great object no doubt to bring the two countries together, but I can-

> not help arguing with myself that if, with the two countries three thousand miles apart, I get so many despatches per week that I can with difficulty attend to them all satisfactorily, what would be my fate if the cable succeeds, and I had to receive and answer them every day? Therefore, I shall wish success to the Submarine Telegraph between Europe and America, but [may it] happen with just about as little delay as may bring it to the moment when I hope to be back in my native country.[33]

Later statements by diplomats gave Adams's remark an air of prescience. In 1871 Britain's ambassador to Russia concurred with a parliamentary committee that "the operation of the telegraphy increased [his] duties, and kept [him] more constantly at work." To be sure, there is reason to doubt that nineteenth-century diplomats often had to exert themselves during periods of normal, noncrisis diplomacy. The historical literature overestimates diplomatic workloads, partly because it tends to focus on moments of crisis, when toil was most feverish. Efforts to construct a more accurate history are hindered by the fact that idle diplomats generally left few records in the archives.[34]

All the same, diplomatic workloads did tend to increase over the course of the nineteenth and twentieth centuries.[35] The use of the telegraph along with other factors—economic development, growing international trade, increasing societal complexity, bureaucratization—contributed to this increased labor. Yet while the onset of telegraphy was only one of many factors intensifying the pressures upon foreign ministries, its particular contributions deserve special attention, especially since they tended to occur during crises when foreign ministries were most susceptible to stress. The telegraph increased the strains upon foreign ministries in three ways: by increasing the temptation to engage in hasty action and micromanagement; by estranging diplomatic officials from local or biological cycles; and by increasing the use of codes and ciphers.

In regard to the first of these, the ability to communicate with distant parts of the world in real time produced great pressure to follow an activist foreign policy. One historian has noted that telegraphy eliminated "the calming effect upon the most excited despatch of lying unread for a month in the darkness of a mailbag." When it was possible to learn of

Telegraph room in the White House, set up to receive news of the Spanish-American War, 1898.

Library of Congress, Prints and Photographs Division, LC-USZ62-90805

and respond rapidly to events occurring in distant parts of the globe, every incident seemed a crisis and every crisis called for a dramatic response. In such circumstances, many statesmen craved action, if only to create an illusion of control that would satisfy public opinion or their own psychological needs.[36] This desire for activity could produce hasty responses to distant, poorly understood events. Leaders with access to the telegraph found it increasingly difficult to adopt laissez-faire policies. In this sense, telegrams begot telegrams, and statesmen required real strength of character to pursue a policy of "masterly inactivity" in the midst of chaotic events.[37]

An examination of British diplomacy before the Crimean War illustrates the way in which the telegraph could tyrannize statesmen and contribute to the adoption of rash and ill-considered measures. On 13 November 1851 operators sent the initial telegrams over the just completed Anglo-French cable. For the first time, Britain was in electric communication with the European continent. The *Times* celebrated this event: "The success of the glorious attempt to unite Great Britain with the various nations of the mainland will exercise a very important influence upon the great political questions of our time. That influence we are proud as Englishmen to think cannot but be thrown on the side of moderation and peace." This prediction would be tested during the Near East crisis of 1852–1854, the first instance in which Britain possessed a telegraphic link to Europe during a serious international controversy.[38] Unfortunately, diplomacy via electric impulse made an inauspicious debut, and this crisis culminated in the most major war among the powers of Europe since the defeat of Napoleon four decades earlier.

The 1852–1854 Near East crisis, which lasted almost two years, may seem an unlikely illustration of the effects of speed upon diplomatic disputes. In retrospect, it seems sluggish compared with later crises—such as those of 1870, 1914, and 1962—which lasted mere days or weeks. Modern commentators have noted the seemingly "slow, even leisurely" pace of events.[39] Yet an examination of British policy during the dispute indicates that the new opportunities for speed offered by telegraphy often resulted in rash, ill-considered actions on the part of statesmen.

Research on the social psychology of crises suggests clues to understanding the effects of the telegraph on British decisionmaking before the Crimean War. Two aspects of telegraphic diplomacy, one subjective and the other objective, made statesmen feel rushed: first, telegrams imparted a sense of immediacy and urgency that gave decisionmakers a subjective sense of being hurried, regardless of the reality of the situation; second, telegraphy accelerated the pace of international crises and consequently reduced the time available for statesmen to reflect. Time constraints, whether external or self-imposed, tend to stress decisionmakers. Stress is not necessarily bad; up to a certain point, performance improves under increasing stress. But beyond that point, performance begins to degrade in a number of ways. For example, high stress generates perceptual and behavioral distortions—including a tendency to ignore unpleasant information, adamant adherence to existing views of the world, a preference for incremental change over rethinking an entire policy in the light of new information, procrastination, and overoptimism—that impair efforts to make good decisions.[40] Moreover, the perception that one lacks enough time to weigh options to avoid an onrushing threat can produce "hypervigilance," a behavior that in its most extreme form is better known as "panic." Hypervigilance can produce a vicious circle: the belief that one lacks adequate time produces stress, which in turn causes one to underestimate the amount of time that one possesses, producing more stress, and so forth.[41] Moreover, such beliefs become a self-fulfilling prophecy as impetuous decisionmaking tends to accelerate the pace of international crises.

Hypervigilant individuals have a predisposition to rush their search process, artificially narrow their options, and, consequently, ignore courses of action more promising than the one they eventually choose. In fact, such individuals frequently respond to danger by taking hastily contrived actions that are *worse* than doing nothing. For this reason, some studies of crisis situations have indicated that last-minute warnings of threats are worse than no warning at all. In addition, hypervigilant actors are predisposed to lose their sense of perspective, neglect the long-range consequences of their actions, suffer from diminished creativity in problem solving, and believe that their own behavior is determined by external forces but that the behavior of others is unconstrained and internally motivated. Of particular relevance to British policy before the Crimean War, hypervigilant individuals often wish to make quick decisions in order to reduce short-term uncertainty. This pushes them toward premature action.[42] Even a brief account of British diplomacy during this crisis offers fertile ground for finding such decisionmaking errors.

For example, the diplomatic circumstances of mid-1853 might have provided an occasion to halt the march toward war. But rather than exploring these opportunities in a reflective and methodical manner, the members of the British cabinet acted with reckless speed. In July, when war between Russia and the Ottoman Empire seemed increasingly probable, representatives of the four European powers not directly involved in the conflict—Britain, France, Austria, and Prussia—met in Vienna to devise a compromise that might conciliate the tsar without depriving the Turks of independence. Britain's foreign secretary, the Earl of Clarendon, hurriedly telegraphed his assent to the resulting document, the Vienna note, which he described as "the means best calculated to effect a speedy and satisfactory solution of the differences." He accepted the agreement without waiting to read a competing Turkish proposal or even to see the Vienna note in its final form. The rapid pace of Clarendon's activities troubled Viscount Stratford de Redcliffe, the British ambassador in Constantinople, who chided his superior for providing instructions that were not "deliberate."

Stratford's concerns were well founded: the Ottoman Empire rejected the Vienna note, thereby exacerbating the crisis. This failure was predictable. The four neutrals, in their rush to achieve a "speedy" agreement, had not adequately considered the views of the Ottoman regime. Constantinople was not connected to the European telegraph grid and, because the sultan did not trust his envoys enough to allow them initiative and flexibility, the Turks had little opportunity to exert prompt diplomatic influence in the capitals of the other European powers. Any fast-paced negotiation conducted outside the Ottoman Empire was almost certain to neglect the sultan's interests and therefore be unpalatable to him. The misconceived nature of this diplomatic effort thwarted the last significant opportunity to avert war.[43]

As war loomed, British politicians became uneasy. On 23 September 1853 Lord John Russell, a member of the cabinet and a former foreign secretary, condemned "the fatal facility of the electric telegraph," which, he believed, was leading British policymakers into dangerous territory. One day later, Clarendon, still serving as foreign secretary, confessed to Russell, "The scramble & pressure of this office at times prevents things being done with all the reflection they require."[44] Russell wrote to Clarendon that in the new age of electric communication events outran diplomacy: "These telegraphic despatches, are the very devil. Formerly Cabinets used to deliberate on a fact & a proposition from foreign Gvts; now, we have only a fact." In its haste, the British government acted without sufficient information. On 23 September 1853 the prime minister and the foreign secretary took the drastic step—Stratford called it "crossing the Rubicon"—of asking Stratford to summon a British naval fleet into the Dardanelles. The two men made their decision on the basis of a single telegraphic report from a foreign government, that of France. They did not consult the rest of the cabinet before making this portentous decision, as they had promised to do. Nor did they wait to receive Stratford's report from Constantinople, which contradicted the French dispatch.[45]

Partly as a result of such hurried activity, which the telegraph facilitated, the peace process collapsed, the conflict escalated, and a war began that, at the beginning of the crisis, none of the political leaders of the involved countries had sought. A few years after these events, Edmund Hammond, the permanent undersecretary at the British Foreign Office, admitted: "I dislike the telegraph very much. In the first place nothing is sufficiently explained by it. It tempts hasty decision." Concise telegrams often conveyed enough data to induce panic without dispensing the information necessary to provide understanding.[46] Decades later, an editorial in London's *Spectator* made the same point:

> It is rumour rather than intelligence which is hurried so breathlessly across continents and seas . . . The constant diffusion of statements in snippets, the constant excitements of feeling unjustified by fact, the constant formation of hasty or erroneous opinions, must in the end, one would think, deteriorate the intelligence of all to whom the telegraph appeals . . . Is it conceivable that statesmen, informed of all they know through such a medium, with its inevitable hurries, its inherent necessity for over-compression, its unavoidable reticences, should ever really *know* even events, should ever be wisely counselled, should ever be internally urged to reflection, as they were under the old *régime?*

Despite the passage of time since the end of the Crimean War, observers still feared the telegraph's incitement to swift action based on inadequate information.[47]

A second way in which the telegraph increased the strain upon diplomatic institutions resulted from the fact that it brought news at all hours and in all seasons, particularly during crises. Whereas twenty-four-hour watches had long existed to guard against "physically present threats," telegraphy forced foreign policy institutions to maintain "around-the-clock readiness" in order "to act on information about far-distant events."[48]

Rapid communication across time zones meant that a diplomatic post might receive most information during its "down time." For example, American diplomats in Europe tended to suffer from the fact that telegrams sent from Washington during normal working hours might reach them at bedtime. Otto von Bismarck, the German chancellor, described the burdensome nature of diplomatic telegrams that required ciphering or deciphering in the middle of the night, and suggested that code clerks be better paid in compensation. The permanent undersecretary at the British Foreign Office remarked, "Telegrams are coming in at all hours of the night . . . nobody can tell the strain upon the mental and bodily capacity of a man from being put upon those things at what may be called unlawful hours."[49] Indeed, the telegraph paid no heed to natural or biological rhythms. As a result, it increased physical, mental, and emotional strain among politicians and bureaucrats. During international crises, such stress contributed to sleeplessness, exhaustion, irritability, and, in the most extreme cases, impaired cognition and emotional disturbance.[50]

Third, because telegrams were easily intercepted, foreign ministries expanded their use of codes and ciphers, creating additional labor. The coding and decoding of messages led to increased workloads and backlogs during crises, when diplomats sent unusually large numbers of long, coded telegrams. For instance, the U.S. embassy in Vienna spent thirty hours encoding a lengthy and important telegram from the Habsburg government during the First World War. During the Paris Peace Conference, German diplomats received instructions to send fewer telegrams and make greater use of the diplomatic pouch in order to reduce the decoding backlog.[51] Decades earlier, Bismarck had bemoaned the added work created by the "encipherment of dispatches—[which] has unfortunately become a habit since the invention of the telegraph." In 1870 a British diplomat agreed with the view that the "use of cipher," which is "necessarily consequent on the habitual use of telegraphing," increased "both the intellectual strain and the mere mechanical labour" of his job. Not until after the First World War did foreign ministries begin to mechanize coding and decoding. Furthermore, the tendency of telegraphers to garble coded messages created additional toil. A parliamentary committee examining the effects of the telegraph suggested that "the omission of a single cipher may . . . cause hours of work to the deciphering clerk." A British ambassador recalled having received telegrams "that took many hours to decipher, and sometimes could not be deciphered at all."[52]

In the case of the United States, concern with secrecy resulted in prohibitions against foreign employees or lower-ranking clerks having access to codes. Elihu Root, the U.S. secretary of state, issued a confidential circular to this effect in 1909. Such policies, when followed, increased the likelihood that those with the highest security clearance would find themselves exhausted by mundane work.[53] One sees here a tension between delegation of responsibility (in order to preserve the energies of top officials for the most important tasks) and the desire to maintain secrecy (which led to burdening top officials with any task considered sensitive). In general, one has the sense that U.S. embassy security was lax, which had the beneficial effect of reducing strain upon diplomats during crises.[54] The British Colonial Office had instituted similar prohibitions against lower-ranking clerks deciphering telegrams. These rules crumbled during the flood of work produced by the Boer War. Lower-division clerks thereupon complained about performing the duties of much better paid upper-division clerks.[55] But regardless of controversies over who coded and decoded the messages, telegraphy increased the workload of foreign policy institutions, particularly during crises.

Foreign ministries, despite their conservatism and attachment to tradition, made efforts to respond to the changes produced by diplomatic telegraphy. Telegraphy gave foreign ministries an impetus to rationalize and modernize their operations. For example, Britain's Foreign Office asked for warning telegrams describing events in progress in order to improve its ability to respond to news in a timely fashion. The U.S. State Department adopted more systematic procedures for handling telegrams after it learned that in several cases outgoing telegrams had not actually been dispatched.[56]

Numbering telegrams was a simple means of establishing a "very necessary control" over the flows of messages. Early on, when foreign ministries rarely sent or received telegrams, numbering them seemed unnecessary. But bad experiences with delayed or crossed telegrams, particularly during important moments, made foreign ministries cognizant of the value of additional controls.[57] In 1878 the British began numbering telegrams in the same way as traditional dispatches. The Auswärtiges Amt, the German foreign ministry, during the years before the First World War, admonished and threatened to fine diplomats who numbered telegrams incorrectly.[58] In 1900 the State Department asked diplomats sending telegrams from abroad to indicate the exact time they gave them to the telegrapher. Yet not until the avalanche of work produced by the

First World War did the U.S. government perceive the wisdom of asking its diplomats to number their telegrams.[59] In this instance, as in many others, the war stimulated foreign ministry reform.[60]

International crises encouraged reform because they increased diplomatic workloads and demanded greater efficiency. Recalling the crisis produced by Austria-Hungary's annexation of Bosnia and Herzegovina in 1908, a member of the British embassy in Vienna acknowledged that "we did occasionally have a rush of work." Returning at this time from duck shooting, he found himself "inundated by telegrams."[61] Such periods of intense toil were particularly challenging after the sustained period of generally amicable relations that followed the Napoleonic wars, which had left foreign ministries unprepared for the greater demands that they would encounter in times of international crisis.

The problem of a sudden increase in work appeared in its most acute form at the outbreak of the First World War. Writing of the war's origins, one historian has noted the difficulty of conveying "the effects of the strain of those long hot summer days on men often reluctantly summoned back from their country houses or the spas where they were spending their holidays." An effort to quantify the volume of diplomatic communication during the crisis indicates that, among the five leading European powers, messages to foreign ministries from officials stationed abroad rose by 351 percent, while messages from foreign ministries to such officials swelled by 359 percent.[62]

The flood of war-related telegrams made prioritizing more difficult just as it became more necessary. In Germany, those charged with deciphering telegrams complained that it was impossible to know which messages to decipher first, since the messages that reached them were "almost without exception" marked as "hurry" *(eil)* or "immediate" *(sofort)*. Noting that he could not be responsible for deciding which urgent telegrams were actually urgent, a member of the cipher bureau suggested that especially pressing messages should receive a third label implying urgency *(dringend)*.[63]

The First World War tested the physical limits of those who operated foreign policy institutions. On 27 July 1914, during the diplomatic crisis that preceded the war, France's ambassador to Russia wrote that he was "overwhelmed with telegrams and callers, my head in a whirl." Before the war, the U.S. State Department used its Paris embassy as a hub from which to forward telegrams to diplomatic posts in Europe and Africa. This system, which had reduced costs, collapsed under the avalanche of war-related work for the Paris embassy. Meanwhile, Charles S. Wilson, the American chargé d'affaires in St. Petersburg, cabled: "Embassy staff . . . entirely unable to properly handle immense amount of Embassy work . . . Hope Department will be able to send assistance." Heavy message traffic strained the eyesight and numbed the minds of code clerks. The chief clerk of the British Foreign Office found that "eight hours was as much as any eyes could stand at a time of this sort of work."[64] The unrelenting wartime toil eventually wore down diplomats. By January 1917, "all the secretaries but one" at the U.S. embassy in Vienna had broken down "through overwork." Similarly, overexertion probably contributed to Woodrow Wilson's increasingly serious medical problems during and after the First World War. The "many hours of many nights" that he devoted to encoding and decoding secret messages to his envoy, Colonel House, could have been more profitably spent, especially since the cipher that he used provided little protection from a capable cryptanalyst.[65]

The strains upon diplomats and bureaucrats sometimes found inappropriate outlets. One gains some idea of the increased stress from which German foreign ministry officials suffered during the war by noting the nervous breakdowns that they inflicted upon the employees at the central telephone office. This problem became serious enough that the German foreign ministry sent out a circular noting that as a result of verbal abuse of the telephone operators, "their nerves give out and they must take a rest, and . . . be replaced by unskilled workers." This resulted in further deterioration of the telephone service and additional outbursts directed at the long-suffering telephone employees. In an effort to break this vicious cycle, an Auswärtiges Amt directive asked foreign ministry officials to display "composure and equanimity when communicating with the telephone operators."[66]

It is difficult to isolate the influence of physical and mental fatigue upon governmental policymaking. It seems plausible that officials suffering from sleep deprivation, shattered health, and frayed nerves might arrive at different decisions than they would have under more favorable circumstances, but this is difficult to measure.[67] The aggregate strains on foreign policy institutions are more easily demonstrated. Such strains were particularly visible during the earliest stages of the crisis when foreign ministries staggered under the sudden increase in their workload.

Before 1914, German officials anticipated the backlog of telegrams that would occur in the event of a European war. Several factors, they believed, would account for congestion during a crisis. They laid particular emphasis upon increased traffic from governmental authorities, the public's desire for information, and an increase in censorship. These pressures would threaten to overwhelm the telegraph authorities and produce substantial delays. To reduce such problems, German officials suggested discouraging the sending of unnecessary telegrams and reducing the length of necessary telegrams; these measures would expedite the conveyance of important, time-sensitive messages.[68]

In the diplomatic crisis of July 1914, the system of telegraphic diplomacy did not operate as successfully as governments might have hoped. The large number of telegrams, many of which arrived late or garbled, strained foreign ministries.[69] At the French foreign ministry on the Quai d'Orsay, the overwhelming flow of telegrams prevented bureaucrats from reading messages carefully during key moments and diverted attention from the most important messages. General mobilization further hindered French diplomacy by calling many officials into service, leaving posts understaffed. Likewise, the great quantity of telegrams that German leaders received encouraged them to ignore those reports that they found unwelcome, such as warnings from their ambassador to Britain advising them not to count on British neutrality.[70] In St. Petersburg, a secretary at the German embassy found that during the crisis "it took an eternity" for diplomatic exchanges to travel between the European capitals, and that "no dispatch ever arrived in Petersburg at the expected time." Meanwhile, the need to decode the large volume of telegrams delayed Russian diplomacy.[71]

The breakdown was even more serious in Britain. On 2 August, two days before Britain declared war on Germany, the German foreign ministry received information that telegrams could no longer be sent from England to Germany. Britain's foreign secretary, Sir Edward Grey, explained: "I am informed that delay has been due to extraordinary congestion. Even our own Government messages have been considerably delayed."[72] Anglo-German service was restored before the British declaration of war, and the potentially serious timing of this breakdown seems to have had no influence on the course of events. Nonetheless, it demonstrates once again that telegraphy often acted as an additional source of confusion, anxiety, and excitement during the period that extended into the First World War. It was for such reasons that a French scholar, writing fifteen years before the July crisis, had declared, "No serious resolution, above all a resolution that could determine peace or war, should be taken on the basis of telegrams."[73]

The international crisis of July–August 1914 illustrates many of the dangers that faster telecommunications media posed to efforts to resolve international disputes peacefully. First, consider public opinion. The crisis unrolled too quickly for the pacifically inclined to mobilize effective opposition to it. Substantial evidence from the time indicates that the prospect of war met with great initial enthusiasm among large segments of the public in all of the major European powers.[74] Yet, from what we know about such crises, it is questionable whether this "rally 'round the flag" effect—which encouraged many people to support their governments as they took a mad lunge toward world war—would have persisted had the crisis lasted a few weeks longer.

Second, telegraphy impaired decisionmaking by bombarding foreign ministries with information, thereby overwhelming them. For some countries, such as Britain and the United States, these pressures produced a temporary breakdown in communications. The timing could hardly have been more dramatic. Following the British declaration of war against Germany, the U.S. ambassador to London noted: "Europe is in the swiftest and most desperate war in history and events happen hourly that we are accustomed to think of as possible only in months. The quickest action everywhere in the world is necessary."[75] Ironically, at the height of this crisis, when telegraphy's potential for quickly resolving international disputes seemed indispensable, it proved unreliable. Delays and garbled messages impeded the diplomats.

Diplomatic Time 5

TELEGRAPHY CHALLENGED the conception of time that existed within foreign policy institutions. Before the First World War, most foreign ministries maintained a genteel atmosphere that accorded poorly with the feverish demands that faster communication technologies sporadically placed upon them. Foreign ministries and diplomatic stations became sites of tension between man and machine. Diplomats who resented modernity warily eyed the telegraph, a potentially revolutionary device that, they sometimes feared, threatened to disrupt their lives. Even diplomats with more accepting attitudes toward technology tended to use the telegraph without fully accommodating themselves to it. The resulting difficulties became most apparent during crises.

A Diplomatic Leisure Class

In order to understand the interaction between technological change and attitudes toward time within foreign policy institutions, one must know something about the culture of diplomacy in the nineteenth and early twentieth centuries. Diplomats belonged to an exclusive fraternity during this era, and they adhered to a protocol built up over the course of centuries. They had far more in common with one another than with most citizens of their own countries. They spoke the same language (French), vacationed at the same spas, had similar cosmopolitan outlooks, and held the same standards of social snobbery.[1]

The individuals who staffed and oversaw diplomatic establishments were attracted by this culture, and they, in turn, helped shape it. What do we know about them? During the era that concluded with the end of the First World War, aristocrats dominated the foreign ministries and diplomatic services of most European countries. This was true of powers such as Austria-Hungary, Russia, and Italy.[2] Germany and Britain, in spite of their vibrant bourgeois classes, followed a similar pattern. During the 1871–1914 period, 377 of 548 German foreign ministry officials were noble, including 56 of 62 ambassadors. This bias raised the ire of many in the German parliament, who ineffectually argued that the narrow social base of the service impaired its adjustment to a changing world.[3] In the British foreign service between 1860 and 1914, 60 percent of diplomats were peers, baronets, or landed gentry, while 24 of the 31 British ambassadors fit into these categories. Just 8 percent of British diplomats in this era came from families associated with commerce or industry. Such statistics support the political reformer John Bright's depiction of the British diplomatic service as "a gigantic system of outdoor relief for the aristocracy of Great Britain."[4]

In France, despite the establishment of the Third Republic in 1870, aristocrats continued to staff many of the most important positions in the diplomatic service. In 1898 nobles headed five of France's ten embassies. As accomplished a diplomat as Paul Cambon, France's ambassador to Britain before and during the First World War, felt insecure about his bourgeois origins.[5] Marcel Proust's famous fictional diplomat, M. de Norpois, aided by "the prestige that attaches to an aristocratic name," was able to hold prominent posts even under left-leaning French governments. For nobles who felt estranged from the state, such as aristocrats in France or princes absorbed into the German Empire, diplomacy was one of the few careers they could honorably pursue.[6]

Matters had begun to change by the twentieth century, at which time France, where training at the École libre des sciences politiques had come to count for more than aristocratic lineage, employed fewer aristocrats and more commoners among its diplomats than did any other European great power. Nonetheless, the aristocracy continued to exercise considerable influence and to provide more than its fair share of entrants. Just as important, cosmopolitan aristocratic mores did much to set the tone of the service.[7] Numerous French diplomats busied themselves with emulating the British nobility, which by the nineteenth century had become the most eminent in Europe. Some strove not just to speak flawless

English but to speak French with an English accent. Such a dialect in a Frenchman suggested that he had been raised by an English nurse. Fluency in the customs and language of the English elite was an authoritative marker of social distinction, especially since the bulk of the French population lacked knowledge of the English language.[8] Writing during the 1940s, a French diplomat vividly depicted this phenomenon when he described his peers at the French foreign ministry, who he felt were "imprisoned by Anglomania": "Naturally the anglomaniacs at the Quai d'Orsay get their clothes at London, smoke . . . an opiated tobacco, and certain of them take matters to the point of absurdity by learning to speak our language with an English accent."[9]

The United States, like Republican France, lacked a legal basis for a hereditary aristocracy. But unlike France, America possessed few vestiges of an aristocratic social structure. The United States is therefore the clearest exception to aristocratic dominance of the diplomatic corps among the great powers. As a result, American diplomats abroad could not compete in the game of social standing, and they sometimes found it difficult to collect political gossip or exert informal influence. Yet U.S. envoys did form an elite, albeit a bourgeois one, and they frequently sought to affect the manners of their European colleagues.[10] The State Department often attempted to operate by the rules that supposedly characterized European "Society." U.S. Secretary of State Henry L. Stimson, concerned with being a good sport, ended his department's involvement in efforts to break the codes of other foreign ministries. As he said, diplomats were "the only class of officers who are supposed to deal internationally on a gentlemen's basis."[11]

The dominance of liberal, capitalist values in American society made public hostility toward U.S. diplomats, with their aristocratic pretensions, inevitable. Some Americans even accused foreign service officers of "going native" and adopting the contempt for the United States typically encountered among the European upper classes.[12] Joseph Grew, an undersecretary at the State Department during the 1920s, declined to testify before Congress because he feared that his aristocratic manners would antagonize the representatives of the public: "[Those] who talk through their nose and spit on the floor will cut a lot more ice than those who talk like Englishmen." Decades later, Senator Joseph McCarthy, in his attacks on the State Department, capitalized on this American populist tradition. In a famous speech he described Secretary of State Dean Acheson as "this pompous diplomat in striped pants, with a phoney British accent."[13]

Whether aristocrats or would-be aristocrats, diplomats disparaged paid labor, especially when confronted with nouveaux riches whose fortunes were larger than their own. Count Georg Herbert zu Münster, Germany's ambassador to France until 1900, complained that his second secretary, who was heir to a great tannery fortune and possessed a title dating only from 1885, "smelled of leather." Another German diplomat, whose family had derived wealth from the trade in champagne, received constant reminders of his background from his peers, who called him "Clicquot" or "Extra-dry." Envoys in the diplomatic service distanced themselves from consular agents, whose association with commerce made them seem grubby and bourgeois. Members of the elite political division at the French foreign ministry snidely referred to the commercial section as the "grocery." Despite reforms of the British foreign service after the First World War, commercial attachés were still "not asked to dine at the Embassy."[14] Likewise, a bias against commercial affairs existed at the U.S. State Department and the German Auswärtiges Amt.[15] An association between Jews and "the taint of trade" contributed to a genteel anti-Semitism that permeated the diplomatic corps in every important capital.[16]

By paying insufficient salaries, foreign ministries demonstrated that their officials served out of noblesse oblige, rather than from a base desire for money.[17] Inadequate pay also served as a useful barrier to exclude those who lacked independent means. Until 1894 entrance into the French diplomatic service required a private income of 6,000 francs. As of 1880, candidates for the German diplomatic service needed to demonstrate a worth of at least 6,000 marks. This figure rose to 10,000 marks in 1900 and 15,000 by 1914. Similar requirements existed in Britain, Italy, Austria-Hungary, and Russia, and low salaries served the same exclusionist purpose in the United States. Even when, as in the case of Germany, public outcry led to the abolition of formal wealth requirements, foreign ministries continued to apply them informally.[18] Overall, as the former chief clerk of the British Foreign Office remarked, "the Diplomatic Service was a natural outlet for capable boys who were not anxious to make money," or, even better, were anxious to demonstrate that they were not anxious to make money.[19]

One of the ways that aristocrats displayed their status and reaffirmed their identity was through the wasteful and conspicuous consumption of time. Such behavior flaunted aristocratic freedom from paid labor. Thorstein Veblen, a contemporary economist and social critic, asserted that the term "leisure" need not "connote indolence or quiescence. What it connotes is non-productive consumption of time." Among the leisure class, time "is consumed non-productively (1) from a sense of the unworthiness of productive work, and (2) as an evidence of pecuniary ability to afford a life of idleness." Likewise, a historian of the eighteenth-century Anglo-American world has noted: "The gentry's distinctiveness came from being independent in a world of dependencies, learned in a world only partially literate, and leisured in a world of laborers . . . Idleness, leisure, or what was best described as *not* exerting oneself for profit, was supposed to be a prerogative of gentlemen only." Hence the popularity of such practices as snail races—for a time a fashionable form of gambling among Viennese aristocrats—which provided a picturesque means of expressing disdain for both time and money. As other classes became increasingly wealthy and cultured, the importance of leisure as a badge of aristocratic identity became all the more important.[20]

According to this logic, diplomats tended to adopt a leisured lifestyle for two reasons. First, the diplomatic corps drew many of its members and much of its tradition from a class that defined itself by its abstention from productive work. Second, while aristocrats avoided undignified labor in order to maintain their own honor, diplomats had an additional obligation to maintain the honor of the king or queen. Contemporary ideals posited that an ambassador or minister functioned as the monarch's personal representative to another court. This notion influenced the diplomatic culture of all members of the Western states system, even republics. It was scandalous for a diplomat to engage in remunerative or undignified labor, as this demonstrated his sovereign's commonness and unfitness to rule. Moreover, the symbolic role of diplomats as servants of the sovereign obligated them to serve as mechanisms for vicarious leisure. As one historian affirms, personal servants in England "added to the dignity of their employer's idleness by being as conspicuously idle as possible themselves."[21] Veblen commented that "the maintenance of servants who produce nothing" allowed a ruler to demonstrate great wealth and social status. Diplomatic displays of vicarious leisure held particular value since the represented sovereign was physically absent and could not conspicuously consume on his or her own behalf. Over time, such traditions gained a cultural inertia that led to their continuation despite changes to the environment in which they operated.[22]

Work Patterns

The aristocratic ethos and the associated emphasis on leisure that permeated foreign ministries hindered the adoption of notions of time discipline identified with the industrial revolution. "Aristocratic time," if it can be contrasted with "bourgeois time," emphasized independence and quality of life rather than punctuality, industriousness, and efficiency.[23] Foreign ministries were a haven for those with aristocratic attitudes toward time. In 1914 the British Foreign Office possessed the aura of an exclusive club, with 40 of the 176 employees engaged as doorkeepers, cleaners, and porters. Studies of Britain's foreign service conducted after the First World War suggested that high social status was more highly prized than initiative. One contemporary scholar referred to the Foreign Office as that "last choice preserve of administration, practised as a sport."[24] More generally, a historian has described the atmosphere prevailing at foreign ministries before the First World War: "When not frequenting the salon, the diplomat or functionary worked at a slow, easy pace in his cosy office with a fire in winter, the finest cigars, a good book, and vintage port. In the late nineteenth century diplomacy was still something of a leisurely hobby and was related to literary pursuits, horse-riding, hunting, and the establishment of fine literary and musical studios."[25]

Foreign ministries displayed considerable inefficiency during the period before the First World War.[26] As institutions, they tended toward extreme conservatism. The British Foreign Office introduced typewriters relatively late, preferring instead to use quill pens. A clerk at the Foreign Office during the late 1890s reminisced about a titanic struggle that resulted from another small effort at reform: "In those days all dispatches were kept folded in the Office, an immensely inconvenient practice. All the other public offices kept them flat, but when it was suggested that the Foreign Office papers should be kept flat, there was a storm of opposition. They had been kept folded for a hundred years; the change was unthinkable." Such attitudes, although a source of frustration for reformers, maintained morale by instilling a corporate identity and a sense of tradition.[27]

General inefficiency aside, foreign ministries occasionally rose to the level of true indolence. For German foreign ministry clerks in Bismarck's time, a noontime appearance at the Auswärtiges Amt, followed by an hour or two of exertion, does not seem to have been an exceptional day. Insiders there declared that many of their peers "spend their time strolling about, and . . . are more often to be found out shooting than in their office."[28] A clerk at the Italian foreign ministry, bored with deciphering and transcribing messages, "decided that it was not worth the trouble to go to the office to sleep when I could sleep more comfortably in my own bed." One of his co-workers, when asked to submit a "half yearly report on some subject appertaining to . . . official duties," composed a sonnet on Zanardelli, the cat that roamed the office. During the early nineteenth century the Russian Imperial Ministry of Foreign Affairs gave sinecures to aspiring writers such as Pushkin. While such extravagance declined over time, a historian has noted that "even in the twentieth century, [Russian] Ministry officials worked at a leisurely pace and had long respites for tea."[29]

Matters were much the same in France and Britain. In France, officials at the political directorship had been required since 1891 to work at least a five-hour day, but "the time of arrival grew ever later," while "the hour of departure advanced." Foreign Minister Delcassé's chef du cabinet conducted a study of the financial division of the Quai d'Orsay, which was notoriously inefficient. On 16 July 1903 he found that only six of the fourteen employees started work within half an hour of when they should have, four were at least an hour late, and one arrived two hours late and then left an hour early. French clerks glorified such conduct, relating the story of a peer who, in response to commentary on his tardiness, answered, "I came late today, but in return, I shall leave early." Some French officials defended the slack rhythm of work, which permitted the young men "to amuse themselves" while the older men had time "to think." Other observers were more critical. Thomas Sanderson, the permanent undersecretary at the British Foreign Office, dismissed the idea that his organization had much to learn from its French counterpart: "I altogether objected to having the French Foreign Office . . . being held up as a model, when its own Chief complained that it was enormously overmanned and yet inefficient, and it was notorious that our Embassy at Paris and others found it difficult to get any ordinary business through it."[30] The famous five o'clock tea at the Quai d'Orsay and the use of copperplate calligraphy, during a time when typewriters were widely used in the business world, harked back to an earlier, less rushed, era.[31]

Similarly, a former clerk reminisced about the relaxed atmosphere at the British Foreign Office, epitomized for him by the good-natured acceptance of his pet bulldog, "who started her diplomatic career there, coming with me every day." This memory embodied for him the lighthearted and "frivolous" mood that prevailed before the First World War: "I feel sure that no modern member of the Foreign Office thinks of bringing his dog with him." For recreation during working hours, the clerks engaged in football and stump cricket in the hallways. The congenial environment attracted shiftless callers to the Foreign Office building: "We had fairly frequent visitors of the *dolce far niente* [pleasantly idle] type."[32]

Nonetheless, foreign ministry bureaucrats considered themselves paragons of industriousness compared with members of their own diplomatic services stationed abroad. For example, British Foreign Office clerks perceived their brethren in the foreign service as slothful social butterflies. One can see why. In 1852 Lord Cowley, Britain's new ambassador in Paris, complained that he could not get any labor out of his attachés before noon. A British diplomat recalled his tenure in Romania as chargé d'affaires before the First World War: "In the ordinary way there is not much work to do: and no one knows quite what to do with the large amount of leisure on his (or her) hands. There must, for instance, surely be a bankruptcy in ideas when one of the daily interests is to take every opportunity of an arrival and departure to go and meet trains . . . and yet this occupation, alternating with desultory tennis parties, becomes part of the regular routine." With little to do, energetic British diplomats wrote verse, plays, or travel books.[33] In the United

States, a number of Boston Brahmins sought diplomatic careers, which they believed would lend them dignity and provide them with enough free time to pursue their literary endeavors. Junior secretaries in U.S. legations struggled to find enough work to occupy themselves during the four hours a day that their offices were open.[34] Their counterparts from other countries often had similar workloads. When William M. Collier, the U.S. minister to Madrid, described his secretary of legation, he captured a persona readily encountered at embassy receptions: "He is good-looking, thoroughly gentlemanly, affable . . . is popular and lives handsomely. His faults, which are not the worst in the world, are the result of his having too much money to make hard work a necessity for him."[35]

Diplomats avidly pursued the pastimes of the aristocracy. In response to inquiries, the German government considered the possibility of granting hunting licenses to foreign diplomats unsatiated by the occasional shooting party. In the United States, most embassies relocated northward during the summer to avoid the oppressive Washington weather. In 1914, for example, the German embassy arranged to move to fashionable Newport, Rhode Island, from mid-June until the end of September. Newport society embraced the German diplomats, who joined the Spouting Rock Beach Association, the Newport Golf Club, the Newport Reading Room, and the Newport Historical Society (which invited them to a "Lawn Fete" on July 13). After returning from the traditional summer in New England, German diplomats—particularly the young bachelors among them—would rejoin their peers from other embassies in "the bar of the Metropolitan Club, at sporting events, and at the countless Washington society parties."[36]

An American diplomat, recalling his days as a private secretary to the U.S. ambassador in London, noted: "In spite of its responsibilities, the English aristocracy lived a life of apparent leisure. There were always companions available with whom I could play tennis, shoot, hunt, amuse myself." In St. Petersburg, the Imperial Yacht Club was a center for sociability, frivolity, and diplomatic gossip; a German envoy established a favorable social position there by losing 200,000 rubles gambling.[37] Some of the secretaries at the British embassy in Paris complained about having to loiter near their offices when they could have been at the races. Diplomats in Vienna participated in the craze for waltzes and bridge parties that swept high society in the Habsburg capital before the First World War. Cultural interests, such as theater and the opera, abounded among these men of breeding.[38] The appeal of actresses and chorus girls often helped sustain these interests, as many diplomats displayed the decadent sexual mores characteristic of the aristocracy. As long as no scandal resulted, foreign ministries showed considerable tolerance for the extramarital affairs—heterosexual or homosexual—of their diplomats.[39]

Diplomats refused to adopt the sharp division between work and leisure that increasingly characterized their industrializing societies. Spas became favored locations for vacations and informal diplomacy.[40] Although Bismarck's favorite spa was Bad Kissingen in Bavaria, the predilection of his emperor forced him to conduct diplomacy from Bad Gastein in the Austrian Alps, which he detested. The spa at Ems acquired fame as a result of a meeting between the French ambassador and the Prussian king in July 1870 that preceded the Franco-Prussian War. Austro-Hungarian diplomats often refused to accept diplomatic postings that were not within striking distance of a reputable spa. Country house weekends provided a similar setting for pleasure mixed with duty. The German and Russian ambassadors to Britain, who were cousins, spent a final weekend together as guests at Lord Lansdowne's country estate during the weekend of 19 July 1914; a few days later Austria-Hungary's ultimatum to Serbia sent Europe into the crisis that would place them on opposite sides in the First World War.[41]

Diplomats' pursuit of conspicuous leisure produced resentment. The rise of anti-aristocratic politicians contributed to an eclipse of diplomatic influence in the United States and Britain around the time of the First World War. While president of Princeton University, Woodrow Wilson battled against aristocratic student societies. He also denigrated the activities of diplomats. In 1905 a young Princetonian desirous of joining the U.S. foreign service asked Wilson for a letter of recommendation. Wilson replied: "There is little of serious importance to do . . . You are too good a man not to choose something that will really be worth stretching your gifts to." As president of the United States, Wilson expressed this disdain by working through his unofficial intermediary, Colonel Edward House, rather than employing normal diplomatic channels. Likewise, David Lloyd George, the prime minister of Great Britain, detested the inefficiency of the Foreign Office and repeatedly circumvented it. "I want no diplomats," he declared in 1917, "diplomats were invented simply to waste time."[42]

Popular clichés about dissipated envoys influenced policy struggles, especially after the spread of middle-class moral standards gave a pejorative edge to such stereotypes. In Germany, rumors of lascivious activity served as weapons that imperialists and militarists wielded with great effectiveness against titled diplomats of influence. As diplomats lost authority, military officers filled the power vacuum, and German foreign policy became more belligerent. Prince Philip zu Eulenburg is the most famous example. Eulenburg led the "Liebenberg Circle," a group of artistically inclined aristocrats—most possessing diplomatic experience—that exercised considerable influence over the kaiser. Eulenburg, who used his influence to impede proposals for foreign adventurism and preventative wars, was criticized for his supposed advocacy of a feeble, effeminate, and pacifist foreign policy. When the liberal imperialist Maximilian Harden publicized accusations of homosexuality among the Liebenberg circle, the scandal destroyed Eulenburg's career.[43] More broadly, these events strengthened preexisting perceptions about the close relationship between aristocratic decadence and homosexuality, an association that served the political interests of both socialist and middle-class politicians. The scandal also reinforced popular notions about the "many diplomats, courtiers, and bearers of crowns" who were homosexuals.[44] In later years, the extreme right in Germany used rumors of sexual scandal to sabotage the positions of influential diplomats whose policies they opposed.[45] In the public mind, cosmopolitanism and the love of peace became associated with sexual aberration.

Even a few diplomats themselves disparaged "a career of almost uninterrupted leisure." Henry Adams, after having assisted his father at the U.S. legation in London during the dramatic years of the Civil War, could not bear the thought of begging for a diplomatic posting "at Vienna or Madrid in order to bore himself doing nothing until the next President should do him the honor to turn him out." Arthur Ponsonby, an idealistic young British envoy, griped in a letter to his sister, "I think it is a horrible feeling one is doing nothing and has no prospect of doing anything." The members of his family felt little sympathy for his grievance. They informed him that he ought to use his extra time to improve himself with edifying books. Leisure time was, after all, one of the advantages to entering a gentleman's profession. Undeterred, Ponsonby, while still a member of the service, published an anonymous article describing the professional trajectory of a British diplomat:

> He may be lucky enough for a year or two to be employed at a post where important negotiations are pending, and where he may probably have three or four hours work a day. But except for this, his services will not be required by his country for more than half an hour, or an hour, a day, as a rule, and very often not at all . . . After twenty-five to thirty years, the Secretary whose career we are following will become a Minister; this does not involve more work; on the contrary, he will probably have still less.

On 17 October 1900 Ponsonby infuriated the permanent undersecretary by presenting him with a memo that denounced the "demoralizing" effects upon young diplomats of "an easy going career of prolonged leisure." Finding his desire for activity frustrated, Ponsonby abandoned a career in diplomacy. He later became a reformer, entered Parliament, and critiqued aristocratic dominance of the Foreign Office.[46]

The Technological Challenge

Aristocrats, like other people faced with a changing environment, sometimes found themselves thwarted by their own values and habits of thought. Veblen described the resulting dangers: "the sense of the shamefulness of manual labour" could sometimes take extreme forms within the leisure class, even overcoming "the instinct of self-preservation." Certainly the notion that work robbed gentlemen of their dignity sometimes proved inappropriate. The story of Jamestown provides a vivid example. An English colony established on the coast of Virginia in 1607, Jamestown eventually opened North America to British settlement. Gentlemen made up about a third of the original colonists. They dreamed of establishing an easy life on aristocratic "plantations" in the New World. As a natural ruling class, they would contribute to the colonial project by serving as military commanders. Unfortunately, as contemporaries observed, to have "more commanders and officers than industrious labourers was not so necessarie." By definition, gentlemen lacked craft skills and did not perform manual labor. Rather, they assiduously cultivated the art of what one historian terms "genteel loafing." To make matters worse, these gentlemen brought along a large number of personal attendants, generally footmen, unskilled in and proscribed from the sorts of occupations—especially growing and gathering food—necessary to survive in the new environment. In May 1611 the governor complained that rather than planting crops the people were at "their daily and usuall workes, bowling in the streetes." Faced with starvation, the colonists sought to acquire food from the Indians (through beggary and theft), engaged in cannibalism, and waited for shipments from England. Many died.[47]

Likewise, aristocratic culture clashed with the precision and efficiency increasingly necessary to function effectively in industrial society. Numerous historians have argued that aristocratic distaste for industrial occupations led to a disdain for formal training in science and technology that may have contributed to Britain's relative economic decline.[48] This anti-technology bias made the assimilation of new technologies more difficult. For example, Britain's Colonial Office resisted the adoption of the telegraph, the typewriter, and the telephone. The Italian foreign ministry possessed a few typewriters, but did not formally hire typists until the First World War. In Germany, prejudice against the technical professions diminished the effectiveness of the imperial navy, since engineers played a crucial role in the functioning of the ships. During the early twentieth century, the German foreign ministry witnessed a struggle over whether or not it should have automobiles available for official use.[49] The German army officer corps, which is often held up as an exemplar of the ability to combine aristocratic values with an enlightened attitude toward scientific progress, displayed suspicion to technology in its resistance to the telegraph and its slow adoption of the telephone. The German army's weakness in the use of telecommunications contributed to its failure at the first battle of the Marne during the First World War. Comparable problems plagued the Austro-Hungarian military in July 1914: the Habsburg government found that a lack of communication equipment prevented it from launching the quick attack on Serbia that, it hoped, would produce a fait accompli and avoid a world war.[50]

When one considers aristocratic distrust of technological change as well as the skepticism with which bureaucracies treat any destabilizing innovation, it is hardly surprising that foreign ministries suffered from technophobia.[51] To the extent that foreign ministries had formal entrance requirements at all, most were biased in favor of law and the humanities, requiring no knowledge of science or technology from prospective employees.[52] This issue became a source of contention in Germany before the First World War as businessmen impugned the ability of the Auswärtiges Amt to function effectively in an age of rapid economic and scientific change. Well into the twentieth century, the U.S. State Department, following in the footsteps of other, more aristocratic foreign services, eschewed technical expertise and professionalism in favor of aristocratic notions of amateurism.[53]

Many of the individuals involved in diplomacy shared a suspicion or dislike of new technologies. Two of Britain's most famous foreign secretaries of the early twentieth century, Sir Edward Grey and Lord Curzon, abhorred the typewriter. One of Grey's friends noted that "he was distrustful of new inventions," and recalled the statesman's antipathy toward the telephone and the automobile. Edmund Hammond, a longtime official at the Foreign Office, protested that the telegraph tended to "make every person in a hurry, and I do not know that with our business it is very desirable that it should be so." The U.S. embassy in St. Petersburg suppressed a low-ranking clerk's suggestion that the use of carbon paper would reduce the amount of transcribing to be done. A historian has explained the failure of this attempted reform this way: "Embassies in those days, with a precariously small amount of required routine work, could scarcely afford to be efficient."[54]

An influential official at the German foreign ministry, Friedrich von Holstein, felt that recent inventions threatened to disrupt his routine: "Among the innovations, which force themselves [on me], is . . . electrical lighting. I do not like it, and have battled against it at the office as well as in my apartment. And even more against the telephone. I would never take an apartment with a telephone. The fear that one can be called from the Wilhelmstrasse would never give me a chance to be in peace. But if I sit here in slippers, I want peace."[55] Among German foreign ministry officials, Holstein was unusually hardworking. Nonetheless, like many other aristocrats, he resisted technologies that endangered his autonomy.

Not all aristocrats loathed new technologies, of course. Some even considered themselves gentleman inventors. Baron P. L. Schilling, a Russian diplomat, made important contributions to the development of electric telegraphy during the 1830s.[56] Lord Salisbury, one of Britain's great foreign secretaries, earned a reputation as "both a statesman and a skilled electrician." He derived great amusement from the catastrophic investigations he conducted into electric lighting on his estate at Hatfield. His daughter wrote that her father's experiments "relieved the monotony of domestic life": "There were evenings when the household had to grope about in semi-darkness, illuminated only by a dim red glow such as comes from a half-extinct fire; there were others when a perilous brilliancy culminated in miniature storms of lightning, ending in complete collapse. One group of lamps after another would blaze and expire in rapid succession, like stars in conflagration, till the rooms were left in pitchy blackness." During one thrilling episode, an assembly of dinner guests had to put out a fire caused by an overheated wire igniting wood paneling near the ceiling.[57]

Despite such excitement, Salisbury's pensiveness and conservatism led him to question whether the discovery of electricity would ultimately prove to be a blessing for the human race. Other, less reflective aristocratic political figures made more casual use of the new capabilities provided by the telegraph and the railroad, which permitted them to spend time at, or even govern from, their country houses. In this way, they used recent inventions to update supposedly obsolescent aristocratic traditions in a rapidly changing world.[58]

Yet technology sometimes carries political and social implications that consumers ignore at their peril; a technological innovation can create problems larger than the ones it was intended to solve.[59] Consider Otto von Bismarck, the German chancellor, who peppered his subordinates with telegrams when he was on one of his estates or at a spa. The telegraph and the railroad allowed the Iron Chancellor to rule Germany while residing at his country manors.[60] But this attempt to eliminate the contradiction between his desire to dominate Germany and his yearning to live the life of a rustic Prussian Junker had a jury-rigged quality that contributed to his ultimate downfall. Bismarck's long spells away from Berlin led him to neglect his duties, buttressed the arguments of those who argued that he was too old to govern Germany, and, most important, made it difficult for him to maintain ascendancy over the new emperor, Wilhelm II. By 1890 Bismarck had lost the emperor's confidence, and Wilhelm forced him to resign soon thereafter.[61] Bismarck's use of the

Viscount Grey of Fallodon (Sir Edward Grey), British foreign secretary from 1905 to 1916.

From Seton Gordon, *Edward Grey of Fallodon and His Birds* (London: Country Life, Ltd., 1937). Image courtesy of the President and Fellows of Harvard College

telegraph is best seen as evidence of his own contradictory and paradoxical nature—as a reactionary modernizer—rather than as proof of the complementarity of aristocratic values and electric telecommunications.

Like Bismarck, Sir Edward Grey, Britain's foreign secretary at the outbreak of the First World War, capitalized upon the new technologies of transportation and communication to indulge a love of nature. Grey sought to spend as much time as possible in the countryside while he held high office. Although in principle unobjectionable, such behavior tended to aggravate the tension between the aristocracy's local, landed interests and the international responsibilities that it had assumed. The public identified Grey with two rural pastimes about which he published, fly-fishing and ornithology. One bird lover advocated the erection of a national monument to the statesman that would depict him grasping a fishing pole in one hand while the other hand offered food to ducks clustered at his feet.[62]

Despite Grey's preference for the natural world over the world of diplomacy, his illustrious family background helped make him an obvious candidate to be foreign secretary in a Liberal government. But some observers disparaged his poor foreign-language skills (unlike Britain's previous foreign secretaries, he addressed ambassadors from other countries in English), his ignorance of the world, and his tendency to speak "of an interview with an Ambassador as of less importance than an appointment with a wild duck."[63] A historian has noted that the prime minister who appointed Grey as foreign secretary considered him to be "a lightweight: untravelled, lethargic, and preferring his birds and rods to the Foreign Office telegrams." David Lloyd George, a colleague in the Liberal party, believed that Grey, having received largely unearned power at an early age, had no understanding of "the hard work" of normal men. Lloyd George further remarked:

> He was the most insular of our statesmen, and knew less of foreigners through contact with them than any other Minister in the Government. He rarely, if ever, crossed the seas. Northumberland was good enough for him, and if he could not get there and needed a change, there was his fishing lodge in Hampshire. This was a weakness—and it was a definite weakness in a Foreign Secretary . . . He had no real understanding of foreigners . . . Moreover, when a Conference in some foreign capital might have saved the situation, his dislike of leaving England stood in the way.[64]

As a result of his provincialism, Grey wrongly believed that other statesmen hated war as much as he did. As he put it in his memoirs, he held in July 1914 "a conviction that a great European war under modern conditions would be a catastrophe for which previous war afforded no precedent . . . I thought this must be obvious to everyone else, as it seemed obvious to me; and that, if once it became apparent that we were on the edge, all the Great Powers would call a halt and recoil from the abyss." Such a view hindered Grey's efforts to anticipate the conduct of the three Central European empires—Germany, Russia, and Austria-Hungary—and lent a passivity to his behavior. Lloyd George recalls that on Saturday, 25 July 1914, in the midst of the week-long international crisis that preceded the First World War, "Sir Edward Grey left for his fishing lodge in Hampshire, and all other Ministers followed his example and left town."[65] The next day Germany's ambassador to Britain, Prince Lichnowsky, who was desperately struggling to avert war, noted the difficulty of doing anything "until tomorrow" since there was "no one to speak to in the Foreign Office."[66] Grey's love of leisure proved disadvantageous during an international crisis. Not only did new technologies accelerate international affairs, they also enabled diplomats to maintain or recover aristocratic habits ill suited to this faster-paced world.

As with telegraphy, other examples of the marriage of aristocracy and advanced technology—such as the railroad or the automobile—although sometimes harmonious on the surface, contained tensions.[67] This makes sense. An anachronistic ruling class with values derived in large part from the preindustrial period is likely to feel uneasy when faced with technologies closely identified with modernity and the development of big business.[68] The new means of communication and transportation that became available during the mid-nineteenth century created countervailing centers of political, economic, and social power that challenged aristocratic dominance. Furthermore, telegraphy, like other mechanical or electric devices, tended to reduce the need for manual workers (such as couriers), and thereby diminished an occupational distinction upon which the very identity of the aristocracy depended.[69]

Technologies such as the telegraph provided a symbol of the economic and social changes that threatened the power of traditional, landed aristocracies. Diplomats resented the diminution of their autonomy within societies increasingly characterized by bureaucracy and professional politicians. Count Khevenhüller, the Habsburg ambassador to Paris, expressed amazement at a critical message he received from a bureaucrat. He angrily declared his refusal to permit "a simple section chief from writing to His Majesty's ambassador in such a tone." Even in the United States, which possessed a more egalitarian social structure than Europe's, one sees parallel developments. Henry Adams, although not a landed aristocrat, was descended from a patrician lineage of presidents, congressmen, and diplomats. If the nineteenth-century United States had a hereditary ruling class, then the Adams family belonged to it. Yet Henry attributed the decline of his family's political fortunes to innovations in communications technology during the 1840s. In particular, he blamed the development of the telegraph, the railroad, and the ocean-worthy steamship: they had "separated [him] forever" from the "eighteenth-century, troglodytic Boston" where he would have been a natural political leader.[70]

The telegraph threatened to rob diplomats not only of their "free time" but of their very sense of time. Many of the landed aristocracy, with their cherished rural estates and outdoor activities, felt stronger ties to natural cycles, particularly the seasons, than did the bourgeoisie, with its roots in the city. Aristocrats, partly because of their greater appreciation of organic rhythms, approached the work they did with a "task orientation" typical of preindustrial labor patterns. To take a prominent example, Bismarck disliked what he perceived as sterile, bureaucratic routine, and let most things slide until he faced a challenge. As chancellor, he oversaw similar labor patterns at the foreign ministry, which lacked regular working hours.[71] But such task orientation appeared wasteful to the bourgeoisie, which attempted to synchronize and accelerate labor by setting it to the clock.[72] By the twentieth century, the German foreign ministry had absorbed some of these more professional attitudes; for example, it sought to dissuade diplomats from believing that "the presence of officials at their posts is . . . only necessary when great, important, and urgent business is at hand or there are matters of a special nature with which to deal." Rather, it sought to inculcate the idea of "steady activity" *(stetige Beschäftigung)*. Such directives should be placed within a broader context of technological change. Telegraphy furthered the effort to inculcate society with bourgeois notions of time.[73] While the railroad frequently receives credit for creating an incentive to align clocks over vast distances, the telegraph, by swiftly communicating the hour, made time standardization practicable.[74] Any discussion of nineteenth-century technologies for imposing time discipline upon labor should consider telegraphy.

Like many other workers, diplomats and bureaucrats did not accept these encroachments upon their temporal culture without a struggle. At times they simply disregarded the new demands. A British diplomat wrote in his diary about an unpleasant surprise that awaited him when he returned from an evening of carousing in Vienna: "I got home, found a note from the Ambassador asking me to come round and decypher a telegram, so trotted round and began it. But it was a whopping great long thing and very dull, so I didn't finish it and went to bed." Similarly, Sir Frederic Rogers, the British Colonial Office undersecretary, described the "presence of mind" that he exercised in order to resist reading a long, seemingly highly important cipher telegram that arrived during dinner and threatened to disrupt dessert.[75]

At other times, opposition to the demands of the telegraph took more overt forms. Edmund Hammond, the permanent undersecretary at the British Foreign Office, opposed installing a telegraph at work: "If they have the wires in, the office should be open night and day." Hammond eventually lost this fight, and the Foreign Office received its own telegraph line. But such concerns had particular resonance with aristocrats, who believed that their freedom from base, mundane concerns made them a natural political ruling class. Their leisured lifestyle provided them with time "for study and reflection and a disposition to take the long view." Telegraphy, by plundering their leisure time and leaving them "too little interval for reflection," seemed to undermine the rationale for their continued leadership.[76]

Such developments seemed all the more intolerable because much of the appeal of a diplomatic career was its association with leisure. As one historian suggests, "Some of the most successful [U.S.] diplomats entered the Service . . . because it was the most socially respectable way they knew of leading a carefree life." According to a circular from 1873, U.S. diplomatic agents received sixty days paid vacation per year, the same amount of time granted to officials at the British Foreign Office and French diplomats.[77] A powerful nobleman such as Count Harry von Arnim could spend several months every year in Germany, although he officially served as the German ambassador in Paris. Despite the importance of the Balkans to Austria-Hungary, Baron Carl Macchio, the Habsburg envoy to Montenegro, retired to Italy during the winter to gather strength for the "summer campaign." Over the course of one year, an Italian ambassador to Greece reportedly spent only fifteen days at his post. Count Philipp zu Eulenburg initially entered the German diplomatic service because he needed ample leisure in which to pursue his artistic interests.[78]

Telegraphy threatened the freedom and the relaxed atmosphere of diplomatic life. One of Bismarck's associates remarked that during critical periods "suffering Legations" received telegrams that "must be answered without delay." This new factor disrupted such traditional diplomatic activities as "shooting parties, visits of 'artist friends' and such-like, musical *matinées,* theatrical rehearsals and . . . those quiet hours, consecrated to the secret improvement of one's mind by aid of French novels." A former British envoy declared that the telegraph prevented many diplomats "from hiding amicably behind their own laziness and inefficiency." Given the high value aristocrats placed on their leisure time, even those who were not generally averse to technology could find themselves at odds with the acceleration of events produced by telegraphy. Diplomats who felt "harassed" by electric telecommunications waxed nostalgic about the days when they had enjoyed an "a-telegraphic paradise."[79]

It would be misleading to portray aristocratic culture as utterly inimical to the successful conduct of foreign policy in the telegraphic era. Overstaffed embassies contained greater resources with which to absorb the increased workloads produced by international crises. Some pleasures furthered the conduct of business. Spas, country houses, and racetracks fostered amicable relations and served as comfortable settings for informal negotiations. Moreover, the self-confidence often characteristic of those born to social leadership imparted an unflappability that frequently proved valuable during moments of crisis. The imperturbable deportment of Lord Lyons and Charles Francis Adams during the *Trent* affair reveals some admirable aspects of aristocratic culture (see Chapter 3). Likewise, Bismarck perhaps owed some of his patience and remarkable sense of timing to his skepticism toward the watch, that object of bourgeois fetishization. "We can set our watches," Bismarck warned one of his ministers, "but time doesn't pass the more quickly because of that, and the ability to wait while conditions mature is a prerequisite of practical politics."[80] Given this attitude, Bismarck could exploit the capabilities of the telegraph without falling victim to the sense of urgency that telegrams conveyed.

Nonetheless, many aristocrats had trouble adjusting to international crises that occurred at an unfamiliar rhythm. During the Russo-Turkish dispute that eventually led to the Crimean War, Viscount Stratford de Redcliffe, Britain's ambassador to Constantinople, sought to refute the assumption that international disputes must occur in accordance with the calendar of aristocratic leisure activities: "You seem to have tumbled into the general mistake . . . You thought that if the Eastern Question could linger beyond 12 August (grouse shooting) it could not by any possibility survive 1 September. The Turks are not sportsmen, as you know, and consequently enter little into such considerations."[81]

The acceleration of international crises in the age of the telegraph made foreign policy officials appear foolish and out of touch. On 3 July 1870 Edmund Hammond, the permanent undersecretary at the British Foreign Office, told Earl Granville, the foreign secretary, "that he had never, during his long experience, known so great a lull in foreign affairs, and that he was not aware of any important question that he (Lord Granville) should have to deal with."[82] This remark gained wide publicity when Granville repeated it before Parliament, and it returned to haunt both men. For that same day a controversy over the succession to the Spanish throne exploded onto the international scene, leading shortly thereafter to a war between France and the North German Confederation.

Because telegrams could arrive at all hours, foreign ministries felt increasingly compelled to maintain an incessant readiness. They consequently expected their embassies to exercise continual vigilance. The telegraph made possible new reporting requirements, even as diplomats continued to pursue amusement. This juxtaposition sometimes produced embarrassment. For example, a sudden change of regime occurred in Romania in June 1930. As it was a beautiful Sunday, most "diplomats found themselves in the country, at the mountains, or on the beach." At the Spanish embassy in Bucharest, no one remained on duty. A Romanian bon vivant, who frequently sponged meals off the Spaniards, feared that his friend the ambassador would be recalled if Madrid was slow to learn of the political revolution. So he took it upon himself to telegraph a message—ostensibly from the ambassador—conveying the news to the Spanish foreign ministry. Aside from any other failures of protocol, the message was written in Romanian rather than Spanish, and it elicited an incredulous response from Madrid. The Spanish ambassador managed to give a satisfactory explanation, however, and the incident became a source of great merriment within the diplomatic corps.[83]

Even officials in highly responsible positions resisted the notion that the capricious winds of international affairs could blow their schedules off course. During the 1911 crisis over control of Morocco, when Britain and France seemed about to go to war with Germany, Winston Churchill, then home secretary, worried about the departure of the first sea lord for a weekend of shooting in Scotland. At a time when war loomed, Churchill learned from the secretary of state for war that "practically everybody of importance & authority is away on his holidays." When an aristocrat did postpone a holiday, the situation was probably very somber indeed. In 1912 the Marquis of San Giuliano, Italy's minister of foreign affairs, expressed shock when Sir Edward Grey, Britain's foreign secretary, cut short a fishing trip to Scotland because of war clouds on the horizon (the First Balkan War broke out soon thereafter): "Grey has interrupted his week-end. This is unimaginable and makes one think that the situation is more serious than I had believed." San Giuliano concluded that Grey must fear the outbreak of a "general conflict" among the great powers of Europe.[84]

German diplomats—particularly but not only the younger ones—frequently deserted their posts. Local responsibilities motivated some of them. Many maintained close connections to the countryside, especially to their ancestral estates. Such attachments produced a sense of alienation from the demands of bureaucratic routine. One count returned to his family estate in Thuringia every year from June until the Oktoberfest to oversee the harvest. Others acted in accordance with the chivalrous and besotted spirit that in 1837 impelled a young Otto von Bismarck to forsake his duties as a Prussian civil servant to gallivant across Europe for three months chasing an Englishwoman. Years later, when Bismarck ruled Germany, he favored an aristocratic brashness in his diplomats. Although he ruthlessly punished insubordination, Bismarck relished underlings who exhibited a well-bred, stylish, and care-free irresponsibility (he used the term "Kavalierperspektive").[85]

Changing expectations made demonstrations of conspicuous leisure less tolerable. In 1896 August Bebel, a Socialist leader in the Reichstag, observed that both bourgeois and socialist critics had expressed concern over the conduct of the German ambassador to Vienna, Count Philipp zu Eulenburg, who had abandoned his post at a significant moment in Balkan politics. Bebel suggested that only sickness or some sort of emergency would have justified a diplomat's absence from Vienna during a "highly important period." But Eulenburg, it appeared, had deserted his duties in order to take a "cruise on a pleasure boat on the North Sea."[86] Likewise, in late 1916, while the German public suffered through wartime austerity, a British periodical published a photograph of the summer activities of the German ambassador to the United States, Count Bernstorff. The snapshot, taken during Bernstorff's visit to the vacation residence of a wealthy New York socialite rumored to be his mistress, showed the diplomat in swimming costume, embracing two ladies similarly attired. It produced an uproar. Bernstorff's many enemies in Berlin—especially military officers and right-wing politicians who wanted to initiate indiscriminate submarine warfare—used the picture to discredit him at a crucial moment in German-American relations. Similarly, in 1918 opponents of the German foreign secretary, Richard von Kühlmann, argued that he could not be performing his job effectively when he wasted so much time on poker and duck hunting.[87]

Increasingly, the German foreign ministry began to consider such apparent acts of unprofessionalism unacceptable. Adopting notions of time discipline characteristic of the bourgeoisie, it sought to enforce more regular labor patterns and rationalize leisure by subordinating it to wider organizational interests. For example, it staggered vacations in order to diminish the loss of efficiency during the late summer.[88] The Auswärtiges Amt also attempted to remove the impression among diplomats and bureaucrats that they had an inviolable right to a vacation, regardless of such factors as the flow of work, what was happening in the world, and the staffing of the office. To this end, the state secretary criticized a diplomat for demanding a holiday. In 1899 the foreign ministry sent a circular to German consular officials complaining of those who exceeded their vacation or were out of contact while they were away. Another circular, from 1909, criticized diplomats for continuing to ignore these rules about vacations.[89] In exchange for demanding greater labor discipline, however, the foreign ministry came under pressure to adopt more professional practices for reimbursing the vacation outlays of diplomats. In the past, while the Auswärtiges Amt had given diplomats considerable vacation time, it had expected them to meet most of their own expenses. The underlying assumption—that diplomats were rich—had led to inadequate salaries and great disparities in the lifestyles of supposed colleagues.[90]

Foreign ministries were often unsuccessful at regulating leisure. Germany's ambassador to Britain, Count Paul von Wolff-Metternich, periodically quit London for weeks at a time without informing his superiors. To create the illusion that he was still at his post, he left behind a stack of papers, blank except for his signature, upon which his assistants wrote reports, which they mailed to Berlin. Such deceptions mattered little so long as significant diplomatic questions did not arise. Crises that occurred during the summer, a time when aristocrats traditionally took vacations, displayed most starkly the clash between accelerating diplomacy and the prerogatives of a leisure class. In such centers of diplomatic activity as London or Vienna, envoys and foreign ministry officials, like the rest of "Society," fled the metropolis for country estates in accordance with "the season." Rumors abounded that poorer members of high society, unable to afford a temporary resettlement to the country, hid in the back rooms of their urban residences during the late summer in order to avoid the obloquy of appearing out of season.[91]

At times, such yearly migrations interfered with the course of international affairs. An assistant to the Russian foreign minister complained bitterly that the summertime absence of policymakers produced long holdups in Russian diplomacy. In 1895, during the first Venezuelan crisis (an Anglo-American diplomatic dispute), summer vacations among officials delayed Britain's response to a quasi-ultimatum from the U.S. secretary of state. This delay probably aggravated the dispute.[92] In the autumn of 1908 a combination of factors—passivity, the departure of key officials on holiday, a casual attitude toward work—among those who managed German foreign policy produced what became known as the *Daily Telegraph* affair. This was a political crisis that occurred when a British newspaper published Kaiser Wilhelm II's intemperate remarks about Anglo-German relations. The chancellor, a former diplomat wedded to his leisure, failed to edit Wilhelm's assertions, and merely passed them on to the foreign ministry. The absence of a number of key officials there ensured that no one else compensated for the chancellor's negligence by vetting the interview. The resulting furor destabilized the German political system.[93]

The historical record contains numerous complaints about vacations during international crises. In the midst of the July 1870 controversy between France and Prussia, a diplomat left in charge of the French embassy in Berlin wrote that "all the members of the diplomatic corps complain . . . of the absence" of anyone with authority to represent the Prussian government. War threatened, but the diplomats who remained in Berlin found no one with whom to speak. They could do nothing more than act as "attentive observers." During the First World War the German foreign ministry investigated possible espionage directed against its embassy in Washington. The ensuing report expressed concern over allegedly lax security during the summer months, a period of the year when most of the embassy staff sought out rustic pleasures, providing little protection for the precious code books.[94]

The July Crisis

The crisis that preceded the First World War is the most important of the summertime diplomatic disputes. It demonstrates how, in the age of the telegraph, the faster pace of such disputes wreaked havoc upon the leisure activities of diplomats. The event that set off the crisis—the assassination of the heir to the Austro-Hungarian Empire—occurred on 28 June 1914, a lovely summer Sunday in the Habsburg lands. The crime coincided with Derby Day, which marked the end of the Viennese social season. Only a few bureaucrats remained at the Austro-Hungarian foreign ministry, and no one of any authority. Count Leopold von Berchtold, the foreign minister, had left on a hunting trip and could not be immediately informed of the tragedy. In Belgrade, an important site for the impending Austro-Serbian conflict, the Russian minister suddenly died. Since much of the rest of his staff had gone on leave, only two diplomats remained to perform the tasks of this crucial post.[95]

The seriousness of the crisis did not become apparent to most observers until the delivery of the Austro-Hungarian ultimatum to Serbia several weeks later, on 23 July 1914. This timing guaranteed chaos, since it fell during the traditional vacation period.[96] On 20 July the London *Times* had commented, "As the London season nears its end the call of the country begins to grow imperious." Shortly thereafter, the U.S. ambassador to Britain left London to spend the rest of the summer in the countryside. Likewise, the British ambassador to Austria-Hungary left Vienna for rural pleasures while the French ambassador to Germany left for a summer holiday in Paris.[97] Germany's representative in Washington and Britain's ambassadors to Paris and Berlin returned home; they had to rush back to their posts when the crisis suddenly intensified. The Serbian prime minister had left Belgrade to campaign for upcoming elections; after this activity, he planned to vacation in Salonika. In Russia,

Serge Sazonov, the tsar's minister for foreign affairs, was at his summer residence in Tsarskoe Selo. The tsar and his family were about to sail for the Finnish Skerries. Prince Troubetskoy, chief of the Near Eastern section of the Russian Foreign Ministry, was at his country seat.[98]

Most such departures made no difference. Yet small, contingent actions, such as the absence of a diplomat from his post, can sometimes have large consequences. This is particularly noticeable during international crises, when history seems at a crossroads. The fact that Austria-Hungary's accomplished but newly installed ambassador to St. Petersburg spent only a few weeks in Russia during the ten months before the outbreak of the First World War surely hindered his ability to collect information about Russian intentions; this may have contributed to his government's disastrous belief that Russia would back down in a serious crisis. Likewise, one wonders whether Russian diplomats might have more successfully challenged the belief in Vienna that Russia, though allied with Serbia, would not intervene in an Austro-Serbian War. The answer is perhaps no; Habsburg decisionmakers were so intent on a localized war with Serbia that they tended to imagine that other powers would act so as to make this possible, refusing to believe evidence to the contrary.[99] Nonetheless, at a moment when good communication was necessary to prevent a tragic misunderstanding, it is worth noting the absence of Russia's ambassadors from their posts at Vienna, Berlin, and Paris. In Vienna, Russia's temporary chargé d'affaires hesitated to act without instructions and therefore failed to articulate the Russian position to the Habsburg leadership as clearly and forcefully as he might have.[100] Would the vacationing ambassador have done a better job? In any case, the willingness of Russian diplomats at important posts to take their vacations at such a moment—despite Russian concerns over Austria-Hungary's anticipated response to the assassination of Franz Ferdinand—says much about the nonchalant attitudes still prevailing among the gentlemen of the diplomatic corps. Although they managed to ignore the new tempo of international affairs for some time, the pace of events eventually overwhelmed them. After seeing their holidays rudely interrupted, they found war virtually unavoidable by the time they regained their posts.

The July Crisis also intruded upon the holiday of the German emperor. On 5 July 1914 Wilhelm II joined the rest of Germany's political leadership in giving support to Austria-Hungary's use of forceful measures against Serbia. The next day, believing there was no prospect of "serious complications," he left for his annual summer trip to Norway, a tradition dating back to 1889.[101] The kaiser's interventions in German foreign policy—associated as they were with erratic judgments and childish outbursts—had frequently proved a hindrance to his country. In this instance, however, his absence probably worsened the crisis. Despite Wilhelm's tendency toward bellicose rhetoric, he ultimately hoped to avoid war between the great powers, much to the consternation of some of his military officers. His former chancellor Bernard von Bülow commented, "William II did not want war. He feared it." But the kaiser was too far away and too out of touch to exercise a direct influence on the crisis during his vacation. He did not return to Germany until 27 July, after he learned that Serbia had mobilized its army. But by then Europe was entering the final stages of the diplomatic crisis, and he failed in his efforts to avert war. A German admiral, Alfred von Tirpitz, later remarked: "In moments which he realized to be critical, [Wilhelm II] proceeded with extraordinary caution. If the Emperor had remained in Berlin, and if the normal Government machinery had been at work, [he] would probably have found ways and means of evading the danger of war."[102]

It is possible that Germany's chancellor, Theobald von Bethmann-Hollweg, purposely encouraged the kaiser's departure. Wilhelm II later reproached Bethmann-Hollweg for keeping him away from Berlin during the crisis. Certainly the chancellor showed no desire to have the emperor return, perhaps hoping that his absence would reduce suspicions among the Entente powers—Britain, France, and Russia—thereby facilitating the localization of an Austro-Serbian War.[103] Austria-Hungary, in an effort to lull the Entente, sent important officials on leave and publicized their departures; it also did not recall military officials and diplomats already on holiday. In this way, Austria-Hungary and Germany sought to exploit the diplomatic calendar and aristocratic notions of leisure to reduce the chance that the Entente powers would intervene when the Habsburgs settled scores with Serbia.[104]

During the July Crisis, political leaders claimed that existing diplomatic practices were too slow to allow for a negotiated settlement in place of war. International relations, conducted at the speed of the telegraph, outpaced the efforts of diplomats. On 29 July the British ambassador reported that the German chancellor "regretted to state that [the] Austro-Hungarian Government had answered that it was too late to act upon [Edward Grey's diplomatic proposal] as events had marched too rapidly." The German ambassador to Russia, when presented on 31 July with a telegram and a letter from the tsar to the kaiser, told the Russian monarch, "I would not know whether letter and telegram would not now already arrive too late." Also on 31 July, the tsar received a telegram from France advising him to delay Russian mobilization (which had already occurred) so as not to give Germany a pretext for mobilization; on it, the tsar wrote, "This telegram has come too late."[105] On 1 August Edward Grey, Winston Churchill, and Herbert Asquith woke their king at 1:30 in the morning and asked him to appeal directly to the tsar urging Russian restraint, but this request was even more tardy than the French one. Writing that same day to the British king, Wilhelm II explained that he could not countermand Germany's mobilization "because I am sorry your telegram came too late." Despite this disappointing message, Grey continued to believe "that it might be possible to secure peace if only a little respite in time could be gained before war is begun by one of the Great Powers." Grey later supported the League of Nations partly because he thought traditional diplomatic institutions had been too slow to stop the rapid escalation toward war during the July Crisis.[106]

There is a self-serving element in some of these declarations about the speed of events. People who have made foolish or sinister decisions often adopt the convenient claim that impersonal forces robbed them of autonomy. Austria-Hungary, Germany, and Russia took important actions that accelerated the pace of the crisis and made a diplomatic resolution of it more difficult.[107] For example, the Habsburg government, hoping to keep the Entente powers from becoming involved in the Austro-Serbian dispute, placed a forty-eight-hour time limit on its ultimatum to Serbia, a remarkably brief period in which to respond to a complex and exacting document. The mere existence of telegraphy and railroads did not require ultimatums to carry such short time limits. To this extent, it is simplistic to see the various political leaders as slaves to their own tools, powerlessly swept toward disaster.

Yet, while telegraphy did not supplant the decisionmaking role of Europe's leaders, it did influence the decisions they made. To take one important example, although diplomatic negotiations require time, and therefore are not ideally suited to a fast-paced technology like the telegraph, telegraphy also facilitates diplomatic intervention, and, in turn, engenders rapid action by those who fear intervention.[108] Telegraphy created the possibility that distant, pacifically inclined statesmen could intercede in the July Crisis. Austria-Hungary wanted to avoid a repetition of the situation in 1912, following the First Balkan War, when rapid telecommunications allowed Britain's Sir Edward Grey to involve himself in the diplomatic settlement. The possibility of intercession impelled the civilian leaders of Austria-Hungary and Germany—both of which hoped for a quick diplomatic victory against the Entente—to seek a fait accompli.[109] These powers exploited the telegraph's potential to accelerate the crisis, and thereby deprived Entente diplomats of the time they needed to resolve the conflict peacefully (that is, via a negotiation that punished Serbia but deprived Austria-Hungary of a clear-cut diplomatic or military victory over the Entente). Telegraphy helped produce an unfortunate situation whereby statesmen with the initiative felt pressured to *act* more quickly because they knew that telegraphy allowed others to *react* more quickly. The resulting acceleration of the pace of events forced everyone to make rushed decisions.

The desire to exploit the advantage of surprise resulted in unexpected, poorly considered, and reckless policies. For example, after Serbia's incomplete acceptance of the Habsburg ultimatum, Austria-Hungary decided to declare war. But how was this to be done? A war declaration is not official until after its reception. Austria-Hungary, having withdrawn its embassy staff from Belgrade, could not announce its belligerence in the normal manner. After Germany refused to deliver the message, the Habsburg government sent it via uncoded telegrams to the Serbian foreign ministry and army headquarters on 28 July. This was the first instance of a government sending telegrams to declare war.[110] The precipitous war declaration made no sense militarily (Austria-Hungary could not actually invade Serbia until about 12 August, two weeks later), proved politically damaging (it demonstrated an unseemly haste to begin fighting), and diminished the chances of a peaceful settlement to the dispute.

Likewise, excessive haste complicated the British declaration of war. A British diplomat noted that the Foreign Office, concerned about the need to respond quickly during a crisis, had devoted "a great deal of attention . . . to the creation of machinery for emergencies in general, and . . . in the event of the outbreak of war, for [the] immediate notification to all British Representatives abroad, high and low, and for automatic telegraphic instructions regarding such necessities as the destruction of cyphers . . . etc., etc. The procedure to be followed had been most carefully worked out, elaborately enshrined in a sort of bible and entrusted for its execution to particularly efficient members of the staff."[111]

The British government prepared to set this machinery in motion on 4 August 1914, after learning that Germany had invaded Belgium. Early that evening the British ambassador in Berlin delivered an ultimatum to the German foreign secretary, warning that his government would break off relations with Germany and act to uphold Belgian neutrality unless shown evidence by midnight (11:00 P.M. in London) that Germany would terminate its invasion. In London, officials waited anxiously as the deadline neared. At 10:40 P.M. (London time), as the British government, having heard nothing from Germany, prepared to release a declaration of war, the Foreign Office rashly concluded that Germany had preemptively declared war on Britain: the Royal Navy had intercepted a wireless message from the German government warning German ships that hostilities with Britain were imminent; British officials incorrectly interpreted this information as meaning that Germany had declared war. On the basis of this unverified report, the Foreign Office quickly rewrote Britain's declaration of war to cite the supposed German announcement, and delivered it to Prince Lichnowsky, Germany's ambassador in London. The Foreign Office soon learned of its error, however, and sent a messenger, embarrassed by his task, to retrieve the war declaration from Lichnowsky and substitute the earlier version. The exchange was made at 11:05 P.M. (London time). A British diplomat later noted that "the Great War, as seen from an official angle . . . opened with [a] blunder." Had the intercepted German message been incorrect and Germany willing to withdraw from Belgium, the blunder would have been more serious, and the British people would have learned that their government had gone to war for fictitious reasons.[112]

During the July Crisis the telegraph accelerated events in a way that, on the whole, served the cause of peace less well than it facilitated the march to battle. While militaries exploited telegraphy's speed to expedite mobilization and present politicians with faits accomplis, diplomats found that it impaired their efforts to slow the crisis enough to avoid world war. There were particular moments—such as the realization by German statesmen that Britain might not remain neutral—when leaders sought alternative policies but, faced with severe time pressures, could arrive at nothing suitable.[113] Such failures were predictable. Individuals who perceive looming danger and strict time constraints tend to suffer from rushed decisionmaking processes, diminished creativity, and poor judgment.[114] The rapid pace of events impaired the efforts of policymakers to resolve the July Crisis peacefully.

At first glance, telegraphy seemed to answer many of the problems of foreign ministries. It provided a means to respond in a timely fashion to the growing demands of public opinion, while simultaneously offering a method for centralizing authority over distant envoys. But despite these new capabilities, many diplomats and foreign ministry officials resisted the telegraph for reasons having to do with a desire for autonomy, technophobia, preindustrial conceptions of time, love of tradition, and a concern that accelerated communication left little time to think and made international conflicts more difficult to resolve. As a result of this resistance, governments tended to use telegraphy without adequately adjusting their behavior to it. Consequently, they often found themselves overtaken by circumstance. The ensuing contradiction—between a disruptive technology and obstinate consumers—became most flagrant during severe diplomatic disputes, such as the July Crisis, when it contributed to instances of chaos and failed diplomacy.

THE MEDIUM III

6 The Zimmermann Telegram

On 19 January 1917, two and a half years after the start of the First World War, the German ambassador to the United States, Count Johann Heinrich von Bernstorff, found himself at the edge of a diplomatic precipice. He already knew that Germany was considering a more aggressive submarine campaign aimed at starving Britain into surrender. On this day, however, he received a secret telegram from his government announcing plans to begin unrestricted submarine warfare (torpedoing all merchant ships without warning in large zones surrounding the allied belligerents) on 1 February. Bernstorff felt certain that this new policy would lead to war with the United States. His superiors shared this concern, as was evidenced by a second telegram he received the same day. This message, which has gone into history as the "Zimmermann telegram" or the "Zimmermann note," sought a German alliance with Mexico and Japan in the event that the United States declared war on Germany. To sweeten the deal for Mexico, Arthur Zimmermann, the German foreign secretary, offered generous financial support and help in reconquering the lost territory of Texas, New Mexico, and Arizona.[1] As directed, Bernstorff's embassy in Washington forwarded this telegram to the German minister in Mexico City.

In his role as the ranking German diplomat in the Americas, Bernstorff had struggled to prevent the United States from entering the First World War on the side of Germany's enemies. Were this to happen, he believed, the enormous economic strength, military potential, and moral influence of the North American colossus would destroy Germany's position as the leading power of Europe. But the control exercised by politically unsophisticated military men over German politics and foreign policy repeatedly undercut Bernstorff's efforts to promote good relations between Berlin and Washington.[2] The German military command supported decisions such as the invasion of Belgium and the submarine campaign against merchant ships supplying the Entente powers, policies that brought only marginal military benefits and produced diplomatic disasters. Strong U.S. economic and cultural bonds with Great Britain—especially among the elite—also hindered Bernstorff's efforts to keep America neutral. Furthermore, many Americans interested in international affairs considered Germany a serious rival to the United States in the competition for trade and political influence in Latin America.[3]

With the beginning of fighting in August 1914, the position of ambassador to the United States became the most important post in the German diplomatic service. Its significance only increased as the war dragged on inconclusively. Germany was fortunate to have someone as capable as Bernstorff to fill it. His background made him well suited to mediate between Germany and the United States. He had spent much of his childhood in London, where his father had been Bismarck's ambassador to Britain. This early experience had fostered his facility with the English language and perhaps also contributed to his strikingly liberal, pro-Western political inclinations. Like many younger sons of the aristocracy, who were unable to inherit entailed estates, he chose a career of government service. Unlike many of his class, however, he cultivated connections with bourgeois and Jewish circles, greatly increasing his effectiveness in a liberal, capitalist society such as the United States.[4] Although accused of laziness—he was once discovered to have plagiarized one of his speeches as ambassador—he possessed a gift for communicating with journalists. His family connections in Germany, which gave him access to the highest levels of the German elite, assisted his effort to serve as an intermediary between the American world, for which he felt great sympathy, and the German world, with which he identified. Early in his ambassadorship in Washington, he received mention as a possible candidate to head the German foreign ministry or even to serve as chancellor.[5]

The war created problems that tested the limits of Bernstorff's aptitudes and connections. Germans expressed frustration with American

Count Johann von Bernstorff, German ambassador to the United States from 1908 until 1917. Mrs. Hugo Reisinger, the woman facing Bernstorff, was accused of running a wireless set for him from the roof of her Fifth Avenue home during the First World War.

Library of Congress, Prints and Photographs Division, Biogr. Files, photo by Bain

criticism of the measures they employed in the name of military exigency (such as the harsh occupation policy in Belgium and submarine warfare). To them U.S. moralism seemed hypocritical: while amassing great wealth by selling supplies to Britain and France, the United States took no effective action against Britain's blockade, which violated international law and was producing starvation in central Europe. In addition, few Germans understood the decisive effect that American belligerence could have on the outcome of the war. How could Bernstorff convince his government of the wisdom of making concessions for the sake of assuaging American sensibilities when German military leaders found the prospect of fighting the United States less distressing than the possible entrance into the war of Denmark or the Netherlands?[6]

Even as the war placed the German-American relationship under stress and increased the psychological distance between the two countries, it impaired the mechanisms for resolving these problems. The British blockade prevented Bernstorff from returning to Germany, and he

felt far removed from policymaking in Berlin. Two and a half years of carnage had produced many changes in his country. The physical difficulties of communicating between Germany and the United States exacerbated the problem of transatlantic understanding. Britain, as one of its first belligerent acts, had located and cut Germany's most important underwater telegraph cables on 4 August 1914.[7] This action, in combination with the Royal Navy's blockade on mail to and from its enemies, largely deprived Germany's foreign ministry of its principal and most effective methods of corresponding with its overseas diplomats: telegraphy and diplomatic pouch.

The German government engaged in innovative efforts to overcome these obstacles, but with only partial success. First, Germany employed wireless telegraphy, still in its early stages of development, to circumvent the British blockade. But atmospheric disturbances frequently disrupted "radiograms." Even when transmission succeeded, Britain and France intercepted and decoded many German messages, although the extent of this practice does not seem to have been fully understood by the German government.[8] Another obvious disadvantage of wireless was that the U.S. government demanded a copy of the code used to transmit all transatlantic radio messages, so that it could censor "unneutral" dispatches (those conveying military information). About letting the United States read coded messages, Bernstorff later wrote: "This course we only agreed to as a last resource as it was not suitable for handling negotiations in which the American Government were concerned."[9] Cargo submarines were a more secure means of transporting messages, but their voyages were too slow and infrequent for satisfactory diplomatic communication in the age of the telegraph. Secret agents who could slip through the British blockade were a theatrical but slow, potentially unreliable, and sporadic way of communicating. And Bernstorff sometimes inserted his own views—concealed by a code understood in Berlin—into seemingly innocuous news dispatches sent across the Atlantic via wireless, a practice that carried significant risks to security and suffered from the unreliability of radio communications. All of these methods, therefore, possessed serious drawbacks.[10]

The cutting of Germany's cables did not entirely deprive Bernstorff and his superiors of access to the telegraph. They made considerable use of Swedish cables. Sweden's long history of antagonism toward Russia engendered a natural sympathy for Germany during the First World War. Although officially neutral, the Swedish government put its telegraph facilities at the service of German diplomats from the early stages of the conflict, including their messages in its own cablegrams.[11] But this route was less than ideal: the Swedish foreign ministry sent the telegrams to Washington via a cable that landed at Britain, allowing British intelligence to monitor them. Fearing that Britain would notice a huge increase in its cable traffic, the Swedish government asked for German moderation in the use of the privilege, and the German foreign ministry tried to limit the number and length of telegrams it sent via this route.[12] Indeed, during the summer of 1915 Britain asked Sweden to desist from carrying German messages. Although Sweden continued this practice, Britain ceased its complaints, apparently because the intercepted messages proved valuable as the skills of British codebreakers improved.[13]

A second telegraphic route soon became available to German diplomats. The controversy that followed the sinking of the British liner *Lusitania* (with the loss of many American lives) in May 1915 convinced U.S. President Woodrow Wilson that a German-American war was possible unless diplomatic methods improved. In a meeting following the disaster, Wilson and Bernstorff agreed that the *Lusitania* crisis and other disputes would be more easily resolved, and future conflicts averted, if the German embassy in Washington could contact Berlin via American cables. The two governments formalized the arrangement after a request by Germany's chancellor to the U.S. ambassador in Berlin, and the United States conveyed the first such telegram for Germany on 2 June 1915.[14] This agreement satisfied both countries: for Germany, it provided another means of transatlantic cable communication; for the United States, it removed one of Germany's excuses for long delays in the negotiations surrounding the *Lusitania* case. After the crisis in German-American relations had passed, Wilson extended the use of American cables to German diplomats involved in peace negotiations to end the First World War.[15] This decision gave German officials a justification for gaining access to U.S. cables during negotiations. As with the Swedish route, U.S. telegrams from Germany passed through the hands of British telegraphers. In both cases, the German foreign ministry believed that its codes offered adequate protection against British espionage.

The deficient state of Germany's overseas communications increased

Bernstorff's autonomy. "Telegraphic communication," he observed, "between the German Government and the Washington Embassy could only be established by devious ways and was thus extraordinarily slow. I had to take decisions on my own responsibility and conduct business with rapidity." He embraced this freedom and frequently put it to good use, as during the *Lusitania* crisis, when "without awaiting instructions from Berlin I exercised my privileges as Ambassador and asked for an audience of the President." Bernstorff believed "that time must be gained," and agreed with Wilson on the sending of an American mission to Germany. "In the meantime," he later recalled, "the exchange of sharp-toned notes between Washington and Berlin went on, without leading to any understanding. But the excitement in the United States gradually died down, and the first crisis was overcome." Bernstorff had helped avert a German-American breach and, possibly, war.[16]

Yet the telegrams that Bernstorff received in January 1917 left him with no room for diplomatic maneuver. Although depressed by the plans for unrestricted submarine warfare, he prepared to implement the policy, even as he quietly struggled to reverse it. He conducted talks with Wilson's advisor E. M. House in the hope of coming to an agreement that would prevent war. On 27 January he sent a desperate telegram—via the State Department—to his superiors in Berlin:

> If U-boat warfare is now begun without further ado, the President will view this as a slap in the face, and war with the United States is unavoidable. The war party here will gain the upper hand, and there will be in my opinion no end of the war in sight, since the power resources of the United States . . . are very great . . . I myself profess the opinion that we will now reach a better peace through conferences than if the United States joins our enemies.

Bernstorff believed Germany would benefit greatly from delaying its intensified submarine war. He contended that if Germany accepted Wilson's peace terms and the Allies rejected them, Wilson would probably not lead the United States into a war on the side of the Allies, even if Germany stepped up its U-boat campaign.[17]

Bernstorff's efforts were fruitless. By January 1917 the governments of Germany and the United States were on a collision course. At 4:10 on the afternoon of 31 January, he went to the State Department to deliver his government's announcement of unrestricted submarine warfare. The American secretary of state, Robert Lansing, noticed that Bernstorff "was not at all the jaunty carefree man-of-the-world he usually was." After presenting the declaration, the normally unflappable envoy had tears in his eyes as he made his departure.[18] Three days later, on 3 February 1917, the United States severed diplomatic relations with Germany, returned Bernstorff's passports, and recalled the U.S. ambassador from Berlin. But war did not immediately follow.

In Germany, few welcomed the prospect of war with the United States. Yet, given the negligible size of the U.S. Army and the massive economic support America was already sending to the Allies, many wondered whether the U.S. entrance into the conflict would make much difference, especially when weighed against the advantages Germany stood to gain from a merciless submarine war against the British. Although no one could be certain that submarines would win the war, the German government decided to gamble upon such a strategy. It hoped that Britain would be starved out of the war—breaking the deadlock on the western front—before the United States could make its presence felt. This sort of reckless decisionmaking had become characteristic of the foreign policy pursued by Germany's military leaders. Military goals trumped diplomatic considerations. Both the kaiser and Germany's civilian chancellor, Theobald von Bethmann Hollweg (who would soon be forced out of office), acted as little more than figureheads.

At the Auswärtiges Amt, the German foreign ministry, a change in leadership indicated the drift of policy. On 22 November 1916, at the behest of Germany's military command, Arthur Zimmermann supplanted Gottlieb von Jagow as head of the ministry. Unlike Jagow, Zimmermann favored an unrestricted submarine campaign against Britain; given the military stalemate in France and Belgium, such a policy seemed the most obvious way to bring about the victory on the western front that Zimmermann fervently desired.[19] Zimmermann hoped the United States would remain neutral under such circumstances, but he accepted the possibility of widened hostilities. To lessen the impact of American belligerence if it did occur, he sought to divert U.S. forces into Mexico (and against Japan), much as he sought to reduce the fighting power of the Entente by supporting insurrections in the British and Russian empires.[20]

Careerist concerns also contributed to Zimmermann's issuance of the famous telegram. He knew something of the Machiavellian politics of Wilhelmine Germany, having worked to ingratiate himself with the military officers who ultimately lobbied for his promotion while they engineered the downfall of his predecessor. Zimmermann's forceful support of unrestricted U-boat warfare had pleased nationalist public opinion and the military leaders. Without such external help, he might never have risen to become foreign secretary. The First World War, which loosened Europe's social structure, assisted the promotion of commoners like him.[21] Whereas Zimmermann's aristocratic predecessor, Jagow, seemed "tired," army leaders believed that someone like the robust Zimmermann—"who could bang his fist on the table"—should oversee the foreign ministry. Such experiences indicated to Zimmermann that his term of office would depend upon maintaining favor with the German military and supporting hard-line policies.[22]

As with many famous blunders, everyone in the German government sought to avoid association with the Zimmermann telegram following Germany's defeat. Historians have noted a "conspiracy of silence" about the telegram in postwar memoirs. Although the evidence is not conclusive, neither the German chancellor nor the German military leaders appear to have known about the telegram before its dispatch (the German military command did support it once it became public). Zimmermann and the kaiser do appear to have discussed in general terms the possibility of a German-Mexican alliance against the United States.[23]

The idea behind the Zimmermann telegram originated with Hans Arthur von Kemnitz, a foreign ministry advisor on Latin American and East Asian affairs. Impressed by the inability of a large force of U.S. troops to capture the Mexican revolutionary leader/brigand Pancho Villa, Kemnitz suggested that Germany entice Mexico into attacking the United States. Such an action, if it immediately followed an American declaration of war on Germany, would reorient U.S. attention and energy away from Europe. Some sort of anti-American agreement with Mexico seemed quite possible since the Mexican government had recently sought Japanese and German support as a counterbalance to the United States. In November 1916 Mexico even offered to provide bases for German submarines. Zimmermann readily agreed to Kemnitz's proposal.[24]

The German foreign ministry sent the Zimmermann telegram to the

CLASS OF SERVICE DESIRED
Fast Day Message
Day Letter
Night Message
Night Letter
Patrons should mark an X opposite the class of service desired; OTHERWISE THE TELEGRAM WILL BE TRANSMITTED AS A FAST DAY MESSAGE.

WESTERN UNION TELEGRAM

NEWCOMB CARLTON, PRESIDENT

Receiver's No.
Check
Time Filed

Send the following telegram, subject to the terms on back hereof, which are hereby agreed to

via Galveston

JAN 19 1917

GERMAN LEGATION

MEXICO CITY

130 13042 13401 8501 115 3528 416 17214 6491 11310
18147 18222 21560 10247 11518 23677 13605 3494 14936
98092 5905 11311 10392 10371 0302 21290 5161 39695
23571 17504 11269 18276 18101 0317 0228 17694 4473
22284 22200 19452 21589 67893 5569 13918 8958 12137
1333 4725 4458 5905 17166 13851 4458 17149 14471 6706
13850 12224 6929 14991 7382 15857 67893 14218 36477
5870 17553 67893 5870 5454 16102 15217 22801 17138
21001 17388 7446 23638 18222 6719 14331 15021 23845
3156 23552 22096 21604 4797 9497 22464 20855 4377
23610 18140 22260 5905 13347 20420 39689 13732 20667
6929 5275 18507 52262 1340 22049 13339 11265 22295
10439 14814 4178 6992 8784 7632 7357 6926 52262 11267
21100 21272 9346 9559 22464 15874 18502 18500 15857
2188 5376 7381 98092 16127 13486 9350 9220 76036 14219
5144 2831 17920 11347 17142 11264 7667 7762 15099 9110
10482 97556 3569 3670

BERNSTORFF.

Charge German Embassy.

The version of the Zimmermann telegram (in code 13040) sent from the German embassy in Washington to the German legation in Mexico City.

National Archives, NWCTC-59-INV15E205-86220212(82A)

Americas via the State Department route (Berlin–Copenhagen–Washington–Mexico City), which crossed British territory.[25] The U.S. State Department (like Sweden's foreign ministry) covertly communicated German messages as received, in this case using the German diplomatic code 0075. An experienced cryptanalyst could tell that German messages differed from the codes normally used by Swedish or American diplomats. It was easy to distinguish between U.S. and German codes, since U.S. codes generally contained letters whereas German codes consisted of numbers.[26] Yet, while identifying the Zimmermann telegram as a German message sent by the State Department was easy enough, actually reading it posed a considerable challenge. Even the staff of the German embassy, which possessed the codebook, experienced some difficulty decoding the telegram because it had been partially garbled in transit. Furthermore, Germany's code 0075 was relatively new and secure, having been in use only since July 1916. The German embassy in Washington had not received the code until the cargo submarine *Deutschland* carried it to the United States in November 1916.[27] After the telegram arrived, the German embassy in Washington recoded it in an older code, 13040, because the German legation in Mexico City lacked the codebook for 0075.

Like most foreign diplomatic traffic transmitted over British cables, the Zimmerman telegram found its way to "Room 40," the center of cryptanalytical activities for British naval intelligence. The two cryptanalysts on duty could immediately identify the telegram as a high-security, presumably important, German diplomatic message. They had trouble understanding it, however, since they had made only partial headway against 0075 when examining intercepted examples of it in recent months.[28] Code 0075 consisted of 10,000 words or phrases randomly represented by numbers ranging from 0000 to 9999. Breaking such a code from scratch usually required identifying the code groups that denoted frequently used words, such as "stop" (period). Although codemakers defended against this type of analysis by using several code groups to represent an especially common word, code clerks tended to memorize only one or two, which they used repeatedly. Codebreakers could often tell when a new clerk had been hired because of the change in equivalents for routinely used words. Knowing the periods permitted codebreakers to divide the message into sentences and to make use of syntactical knowledge (for example, German grammar tends to place a verb right before the period). They could also look for the customary phrases of which diplomats are enamored. In this manner, they began to decipher the message.[29]

Unable at first to understand its full meaning, the British codebreakers uncovered the phrases "unrestricted submarine warfare," "war with the U.S.A.," and "propose . . . an alliance." The telegram seemed to be seeking a coalition against the United States in anticipation of war. After making this initial progress with the intercept, Nigel de Grey, a young codebreaker, called it to the attention of Captain William Reginald Hall, the director of naval intelligence, asking him, "D'you want to bring America into the war?" Hall had commanded a battle cruiser until poor health had forced him into less strenuous activities with naval intelligence. His temperament suited the spying profession, with its deceptions and mind games. His taste for fanciful stories continues to plague the historical record. While their chief thought about how to use the information, the codebreakers went back to work, aided by additional intercepts in 0075 of Bernstorff's telegrams about peace negotiations and arguments against unrestricted U-boat warfare.[30]

As is often the case in the world of espionage, Hall found himself facing the thorny problem of how to use the information that his organization had acquired. At one point he told the codebreakers, "You boys think you do a very difficult job, but don't forget I have made use of the intelligence you give me and that's more difficult." He recognized the immense propaganda value of the intercepted message, which provided a means to bring America into the war. Yet how could he release the information without alerting the Germans to the insecurity of their diplomatic communications, thereby making it much more difficult to collect future intelligence? Furthermore, the usefulness of the Zimmermann note depended upon the perception of its authenticity, but how could the British establish its authenticity without revealing the source of their information?[31]

In addition to concealing Britain's success at breaking German codes, Hall wanted to keep the American government ignorant of Britain's practice of intercepting and decrypting U.S. telegrams. American knowledge of British espionage practices would have hindered efforts to read American messages and, more important, antagonized the U.S. government and public. In the short run, Hall decided to wait and see if the United States would declare war following Germany's announcement of unrestricted submarine warfare. This delay also allowed his codebreakers more time to untangle the message, which was crucial since a partially deciphered message might create a completely erroneous impression (the unknown words might negate or change the meaning of everything else) and would make obvious to Germany that Britain had acquired the message through cryptanalysis.

Hall solved these problems by arranging to steal a copy of the version of the Zimmermann telegram sent from Washington to Mexico City in code 13040. This was easily done since a British agent had access to telegrams that passed over Mexican lines. There were insignificant but obvious differences between the version of the telegram that Bernstorff received and the one he forwarded (date, addressee, the request to forward it). If the U.S. government released a telegram intercepted between Washington and Mexico City, Britain would seem completely unconnected to the event and could maintain the fiction that German and American telegrams were safe from British signals intelligence.[32]

Hall arranged to tell Walter Hines Page, the U.S. ambassador in London, that Britain had "obtained possession of a copy of the German cipher code used in the [Zimmermann note] and have made it their business to obtain copies of Bernstorff's cipher telegrams to Mexico, amongst others, which are sent back to London and deciphered there . . . The copies of this and other telegrams were not obtained in Washington but were bought in Mexico."[33] All of these statements were more or less true, but misleading. Stealing the message from Mexico City allowed the British to use a cover story that was plausible and, in a sense, accurate.

To give credibility to his claims about the Zimmermann telegram, Hall gave the Americans information that would enable them to find it in the records of the Western Union Telegraph Company (its date, the number groups at the beginning and end). He also gave them a copy of the coded text and the decode into German. Britain employed a trusted messenger to reduce American skepticism: A. J. Balfour, British foreign secretary and a former prime minister, who enjoyed the respect of many Americans. The U.S. secretary of state, Robert Lansing, declared, "The fact that Mr. Balfour himself conveyed this information . . . convinced me that it was genuine."[34]

In order to protect his organization's methods of collecting intelligence, Hall planned a misinformation campaign timed to coincide with the publication of the Zimmermann telegram. He would counteract German suspicion by encouraging underestimation of British intelligence capabilities. After the release of the Zimmermann dispatch, he planted an article in London's *Daily Mail* that harshly criticized British intelligence for being outperformed by the United States in this case. He also encouraged the American press to believe that U.S. agents had secretly acquired the telegram, possibly by breaking into a mysterious (perhaps Swedish) trunk, containing diplomatic documents, carried on the same ship that took Bernstorff back to Europe.[35] Further rumors, and variations on rumors, abounded: Britain had acquired the telegram by bribing Dutch postal workers; a discontented German agent had burgled Bernstorff's safe in Washington; a German official at the Washington embassy had sold the message and other documents to the Americans; American soldiers at the Mexican border had captured a German courier carrying the message; a copy had been taken from Bernstorff at Halifax. Robert Lansing facilitated Hall's work by declaring that he could not reveal how the U.S. government had obtained the telegram since this might endanger someone's life. He thus implied that American intelligence had obtained it through either bribery or theft, presumably in the United States or Mexico.[36]

With his preparations complete, Hall set events in motion. On 24 February 1917 Ambassador Page reported to Lansing that Balfour had given him an English translation of the Zimmermann note. Balfour, who already had a long and illustrious career behind him, later called this "as dramatic a moment as I remember in all my life."[37] Page wired the note and some explanatory material to the State Department, where it arrived at 8:30 on the evening of 24 February. The length of the message made the decoding process time consuming. The acting secretary of state, Frank Polk, took it to Woodrow Wilson the next day at 6:00 P.M. The president initially reacted with indignation, and Polk had to dissuade him from publishing the telegram immediately.[38]

Lansing returned from a trip on 27 February and met with the president. Wilson, after ruminating upon the telegram, felt some doubts about it. How could it be authentic when Germany had no means of sending telegrams to America? Lansing explained that Zimmermann had probably used the facilities offered to him by the United States for peace negotiations. According to Lansing: "The President two or three times during the recital of the foregoing exclaimed 'Good Lord,' and when I had finished said he believed that the deduction as to how Bernstorff received his orders was correct. He showed much resentment at the German Government for having imposed upon our kindness in this way and for having made us the innocent agents to advance a conspiracy against this country." Wilson and Lansing agreed on the wisdom of publishing the note. On the evening of 28 February 1917 Lansing paraphrased the Zimmermann note for a representative of the Associated Press. It hit the front pages of American newspapers the next day and occasioned a profusion of patriotic speeches by members of Congress. By a vote of 403–13, the House of Representatives passed the president's bill to arm U.S. merchant ships. Some questioned the note's authenticity, however, and the State Department asked that an American official in London be allowed to decipher the telegram acquired from Western Union so the U.S. government could say it had obtained the message from its own people.[39] Both the U.S. and British governments felt great relief when Zimmermann ended all doubts and acknowledged the telegram's authenticity.[40]

No serious study of the Zimmermann dispatch holds it primarily responsible for the U.S. declaration of war against Germany in April 1917. Wilson's war message to Congress mentioned the telegram but did not give it a particularly prominent place in the speech. The German resumption of unrestricted submarine warfare was the immediate cause of the declaration, while strategic considerations, economic interests, and a concern with national prestige provided underlying motivations.[41] Nonetheless, the publication of the telegram did significantly influence American attitudes. The telegram, and Zimmermann's brazen avowal of its authenticity, highlighted the German challenge to American interests in Latin America, raised security concerns, and demonstrated that neither America's threats nor its goodwill meant very much to the German government. Although the United States would probably have entered the First World War without it, the Zimmermann telegram increased American war fervor. Two of its effects were especially important: it angered President Wilson, and it largely eliminated public opposition to a war against Germany.

If human emotions influence international politics, and they surely do, Zimmermann's telegram had an important effect on German-American relations. Wilson's top advisors, E. M. House and Robert Lansing, already believed the United States should declare war on Germany. The telegram did not significantly alter their views.[42] But Wilson felt betrayed when he learned that Zimmermann had used an American cable, made available to the German government for the purpose of peace talks, to bargain away U.S. territory and seek an anti-American alliance. Wilson's well-known sensitivity to attacks or slights guaranteed that he would take this incident personally, and that he would feel considerable animosity toward the German government. Wilson's reaction deserves especial attention because of his assertive management of American foreign relations and his frequently solitary style of decisionmaking.[43] More than most U.S. presidents, he shaped his country's foreign policy.

The Zimmermann telegram did not prompt Wilson to take dramatic action. He had already resolved to ask Congress for authority to arm U.S. merchant ships, and he would not decide upon the necessity of war for some weeks. But it did destroy his faith in the possibility of working with the German government. On 28 February 1917, the day before the publication of the telegram, Wilson met with a delegation of peace activists. One member of the delegation, William Isaac Hull, a prominent Quaker, later wrote: "President Wilson . . . stressed repeatedly his conviction that it was impossible to deal further in peaceful method with [the German] government . . . I recall with great vividness his tone and manner—a mixture of great indignation and determination—when he said: 'Dr. Hull, if you knew what I know at this present moment, and what you will see reported in tomorrow morning's newspapers, you would not ask me to attempt further peaceful dealings with the Germans.'" Jane Addams, the acclaimed social reformer, attended the same meeting and reported: "The President's mood was stern and far from the scholar's detachment as he told us of recent disclosures of German machinations in Mexico and announced the impossibility of any form of adjudication." By ending his faith in German-American diplomacy, the Zimmermann note had shattered Wilson's greatest hope: that he would be able to negotiate a peaceful end to the war in Europe.[44]

The Zimmermann telegram's effect upon public opinion in the United States was obvious to contemporary observers. Consider the situation before its release. On 31 January 1917, when Germany suddenly announced a policy of unrestricted submarine warfare, the *New York Times* reported that U.S. officials "do not conceal their disquietude over the mental unpreparedness of the American public" for a possible war with Germany. Such fear contributed to the eagerness of Wilson, House, and Lansing to see the note published. A few days after its publication, Lansing observed with satisfaction that the American people "are extremely enraged at the perfidy of the German Government in talking peace and friendship with this country and at the same time plotting a hostile coalition against it." He thought the telegram's release had a greater effect upon public opinion than had Germany's announcement of unrestricted submarine warfare.[45]

The confluence of important events and the lack of polling data make it impossible to speak with certainty about the causes of shifts in American public opinion during the early part of 1917. Nevertheless, research indicates that "a large number of editors in all sections [of the country] called for war for the first time." It also reveals a collapse of publicly expressed pro-German sentiment.[46] While these changes in public opinion after the telegram's publication did not in and of themselves push the United States into war, they did largely eliminate public resistance to a war with Germany.

In Germany, the telegram's publication had much less impact. Zimmermann's popularity with his countrymen did not appear to suffer as a result of his blunder. Kurt Riezler, a confidant of Chancellor Bethmann Hollweg, labeled the proposal of an alliance with Mexico "stupid nonsense" *(Blödsinn)* and "amateurish" *(dilettantenhaft);* he called Kemnitz, who first proposed the alliance, an "incredible idiot" and castigated Zimmermann for listening to him. But Riezler made these comments in his private diary, and they had no effect on the public debate. The socialists publicly criticized Zimmermann, and one of their members in the Reichstag alleged, "Washington will not forget this duplicity for a long time." Nonsocialist newspapers and politicians, in contrast, generally praised Zimmermann for his energy, although some deplored the proposal's publication. Many of the more conservative voices suggested that the public revelation of the note would probably prove beneficial since it might deter the United States from declaring war on Germany. After hearing Zimmermann's explanation, a committee of the Reichstag expressed its approval of his action, at least in principle.[47]

Ironically, Bernstorff's popularity suffered much more than Zimmermann's. The former ambassador's friendly farewell at New York displeased many Germans, who resented his popularity in America. He then suffered harassment from customs authorities in Halifax, who delayed his departure from North America for almost two weeks. Bernstorff first learned of the publication of the Zimmermann note on 2 March, while he was still on the Danish ship that was ferrying him and his embassy staff from America to Europe. The *New York World* sent him a radiogram asking him to comment on the Zimmermann controversy. Bernstorff ignored this request but did respond to queries from the German foreign ministry about the security failure. He denied that the breach had occurred in the Washington embassy.[48]

Zimmermann had a strong interest in making it appear that a traitor had sabotaged the attempted alliance with Mexico. On 2 March the German news service Wolff articulated Zimmermann's viewpoint and defended his diplomacy, saying that the "Imperial Government not only had the right but also the duty to take steps in view of an armed conflict with the United States to compensate for the entrance of new enemies on the scene." The alliance proposal, otherwise well conceived, had been a victim of treachery. The news dispatch further declared: "It is unknown how knowledge came out of this secret communication. However, the treason, for it is certainly that, seems to have taken place on American territory."[49] This precipitate suggestion made Bernstorff and Germany's minister to Mexico the two leading suspects responsible for the failure.

Zimmermann spoke before the Finance Committee of the Reichstag and maliciously attempted to "defend" Bernstorff by raising groundless accusations against him: "How the indiscretion was committed, I can today still not say. I cannot imagine that the Imperial Ambassador, as I read in a newspaper story yesterday, gave the instruction to his valet in order that he might deliver it to Mexico. I cannot believe that Count Bernstorff has acted so carelessly." Such statements should be placed within the context of a campaign that much of Germany's military leadership and various right-wing politicians had already been waging against Bernstorff, whom they scorned for his liberalism, his opposition to indiscriminate submarine warfare, and his desire for a diplomatic resolution to the war.[50]

Once Bernstorff and the rest of the staff of the Washington embassy had returned to Berlin, Zimmermann commissioned a privy councilor named Goeppert to investigate the security breach. Bernstorff's subordinates strongly supported their chief's contention that the lapse had not occurred within the German embassy in Washington. Nonetheless,

Goeppert was deceived by Hall's subterfuges. His report did not even discuss the possibility that Britain might have been involved in the interception of the telegram. Assuming that the United States acquired the message through its own intelligence service, the report argued that German codes must be secure; otherwise the announcement of unrestricted submarine warfare would not have surprised the United States. The report gave considerable weight to Lansing's deceptive comment that he could not reveal how the U.S. government had obtained Zimmermann's dispatch because to do so would endanger someone's life. This implied that someone with inside access to the telegram had stolen it in an act of treachery.[51]

The report's conclusion that some person had probably given away the content of Zimmermann's message, rather than the code, led Goeppert to note that many more people had handled the message at the Washington embassy than at the Mexico City legation. This suggested that the betrayal had occurred under Bernstorff's watch, and placed ultimate responsibility upon him. The true explanation—that a foreign intelligence agency had intercepted the message before it reached Washington—was never seriously considered. Goeppert's report ignored rumors that Britain possessed knowledge of German codes. Bernstorff had repeatedly asserted this explanation, and even Zimmermann privately admitted it as a possibility.[52]

It is probably more than a coincidence that Goeppert's conclusions accorded with Zimmermann's political interests. Investigations into disasters are frequently designed to protect the interests of powerful actors.[53] Bernstorff and Zimmermann had quarreled over policy matters such as the wisdom of pursuing a peace without victory (their views particularly clashing in regard to the western front) and the merits of unrestricted submarine warfare. Zimmermann appointed Goeppert to examine the security failure, and must have been heartened by the ensuing report, which tended to discredit his opponent. A concurrent whispering campaign against Bernstorff undermined his standing with the kaiser. Not until nearly two months after Bernstorff's return did the emperor agree to meet with him, an extraordinary delay for an ambassador returning from such an important post in wartime. Part of the reason for this delay was the kaiser's belief that Bernstorff was responsible for major security failures, especially the publication of the Zimmermann telegram.[54] The rumors promulgated by Hall and Zimmermann proved quite successful at damaging Bernstorff's political influence in Germany—but the hegemony of the military clique over German foreign policy would have stymied Bernstorff's peacemaking efforts in any case.

The end of the war did not greatly improve Bernstorff's political fortunes. He gained little benefit from having opposed and predicted the failure of the disastrous submarine campaign. In his own lifetime (he died in October 1939), his major political efforts were marked by futility. He became a strong supporter of the League of Nations as well as an outspoken critic of National Socialism and anti-Semitism.[55] These were not, alas, profitable stances for a German to take during the 1930s. Hitler's ascension to power turned Bernstorff into a refugee. He died in exile, unmourned and largely forgotten by the country he had so ably served.

What does the story of the Zimmermann telegram indicate about telegraphic diplomacy? Three observations seem pertinent. First, it displays the vulnerability of telegraphy to security failures. Second, it demonstrates the difficulty of using the resulting signals intelligence. Third, it highlights the problems of engaging in secret diplomacy in an increasingly transparent world.

First, the issue of security: telegrams were more vulnerable to interception than were messages carried by diplomatic pouch or courier. If the German submarine *Deutschland* had carried Zimmermann's message to the United States, as he had originally planned, his proposal would have remained obscure until its discovery by a historian many decades later. But the *Deutschland* required a month to cross the ocean, whereas Zimmermann wanted the dispatches to reach America before the announcement of unrestricted submarine warfare on 1 February 1917, less than two weeks away. As Zimmermann later noted, "no time could be lost." Therefore the German foreign ministry instead sent the Zimmermann telegram to Washington via telegraph. Lacking the protection provided by trustworthy couriers, diplomatic telegrams (which were routinely handled by foreign bureaucrats) depended almost entirely upon encryption to preserve the secrecy of their contents. In retrospect, the quality of existing diplomatic codes made the possibility of a security breach obvious and predictable. In Bernstorff's words, "Nowadays there is no cipher which is absolutely safe, if it has been in use for some time."[56]

This incident also exposes some security errors repeatedly committed by the German government during the First World War. The Auswärtiges Amt sent the same message in two codes (0075 and 13040). Such sloppiness is a "capital crime" in the world of cryptography. Britain obtained both versions and compared them, a strategy it often used to break German codes. British cryptanalysts encountered considerable difficulty with the version in 0075, and were able to establish a perfect cleartext copy (good enough to make the German government believe that British agents had swiped a decoded transcription) only because they could compare it with the version in 13040, a code with which they had long been familiar. The message in 13040 therefore provided Germany's enemies with a Rosetta stone that considerably damaged the future utility of 0075.[57]

Moreover, in the case of the Zimmermann telegram, the German government sent a highly sensitive message in an old code. Code 13040 was so out of date that words such as "automobile" and "U-boat"—extremely prominent by 1917—were not in the main body of the vocabulary and had been added as supplements. The Auswärtiges Amt had first used code 13040 in the Americas during the 1907–1909 period. The continued use of a code makes codebreaking much easier. Hall later testified, "Our cipher experts were able to decipher the German ciphers wherever . . . a large number of different messages in the same cipher were available for study and comparison." In its defense, the Auswärtiges Amt felt it had little choice but to continue using 13040. The British blockade hindered efforts to send a new code to the German legation in Mexico City. Nonetheless, if the German foreign ministry needed to continue using this code, it should have taken account of its possible insecurity.[58]

The Auswärtiges Amt committed another serious error: it did not change codes after a security failure.[59] Expense, time, and labor provided powerful disincentives to continually creating new codes. In the midst of a great war, other priorities seemed more pressing. Nonetheless, after the publication of the Zimmermann telegram, the continued use of code 0075 was difficult to defend.[60] A German diplomat wrote, about a year after the interception of the telegram, that there was probably no reason to be concerned if the Entente powers obtained copies of German telegrams since the foreign ministry's best codes, including 0075, "are generally regarded as unbreakable."[61] This statement was doubly foolish because, even if Britain and the United States had not previously broken 0075, they could now obtain a copy of the coded telegram from their own telegraph services to compare against the plain text version that had appeared in newspapers all over the world. The publication of a formerly coded message in cleartext compromises its encryption system, and the continued use of a code in such circumstances displays recklessness.

These principles may seem obvious to modern observers, but such professionalism was foreign to the world of diplomacy, typified as it was by gentility and amateurism. As I discuss in the next chapter, other governments had as many problems with security as did Germany. By 1917 foreign ministries had used the electric telegraph for more than half a century. Nevertheless, they were still hammering out proper security procedures, and doing so under the pressure and adverse conditions produced by the First World War.

The second observation relates to the challenge of using diplomatic signals intelligence. This provides a cautionary lesson, which should temper the tendency to treat the story of the Zimmermann telegram as a simple parable demonstrating the danger of lax security and the value of effective intelligence agencies.[62] Consider the heroic efforts of British intelligence to break the German code, followed by its even more ingenious stratagems to obscure the breakthrough. Such accomplishments tend to distract attention from the connivance of the two governments—Germany and the United States—whose security Britain violated. Both inadvertently cooperated to create the legends that have surrounded the event. In so doing, they helped perpetuate their own vulnerability to British signals intelligence.

As I have argued, Zimmermann misdirected the investigation of the security breach in the hope of exonerating himself and incriminating Bernstorff. In doing so, he prevented the ministry he directed from discovering the insecurity of its communications. Likewise, the U.S. government sought to conceal President Wilson's naïveté in sending ciphered telegrams on behalf of the German government. In addition to looking foolish, such behavior on behalf of Germany violated American neutrality. In response, the State Department authorized the false statement that "no messages ever were transmitted for Germany to or from Berlin through the department without knowledge of their contents." Political and bureaucratic interests led the German and U.S. governments to delude themselves into thinking that their security arrangements kept their diplomatic traffic secure. Both governments participated in their own victimization, facilitating Hall's efforts to fool them.[63]

The story of the Zimmermann telegram demonstrates the difficulty of using signals intelligence. Very soon after beginning work on the Zimmermann telegram, the British codebreakers believed that it might change the course of the war. Yet the British government used the telegram with extreme caution, and only after extensive preparation. In part, this hesitancy resulted from a legitimate desire to preserve British signals intelligence for its most vital mission: keeping the German surface fleet under surveillance and unable to mount a surprise attack. But caution also ensued from the tendency of intelligence agencies to mistrust politicians, a suspicion that can undermine democracy and has often led to the squandering of useful intelligence.[64] Intelligence agencies are often far more concerned about protecting their ability to collect information than about ensuring that the information is effectively used. What started as a means becomes an end, and the aim of their endeavor is forgotten.

Third, the events surrounding the Zimmermann telegram demonstrated the difficulty of employing diplomatic methods that required secrecy to a world in which such trends as faster communication, a freer press, and more politically informed publics were making international relations more transparent. As communication within and between societies became easier, those who sought to control information faced greater challenges. Given these changes, there was a realistic as well as an idealistic aspect to Woodrow Wilson's call for a more open system of international relations. In addition, as mentioned earlier, telegraphy aided espionage. Even during earlier years, diplomats were well advised to "write nothing that could not be read by all of the world."[65] In the age of electric telecommunications, actors needed to make even more allowance for the possibility of leaks and security failures. One can make an instructive comparison between Zimmermann's telegram to Mexico and Page's telegram reporting on it to the State Department. Page's telegram, conveying information from Hall to the U.S. government, was misleading but largely truthful. Hall had gone to great lengths to avoid actually lying to Page and the Americans. Furthermore, had Page's telegram become public it would have done no harm, which made it well suited for diplomatic communication between two relatively open societies.

The publication of Zimmermann's telegram, in contrast, embarrassed many countries, especially the United States, Japan, and Germany. All had reason for annoyance with Zimmermann. The telegram made the U.S. government seem gullible for unwittingly transmitting plans for an anti-American alliance. It also threatened to damage Japanese relations with the Entente powers at a time when Japan hoped to make gains in China. For this reason, Japanese diplomats publicly characterized Germany's alliance plans as ridiculous, while Japan's foreign ministry suggested that "mental delusion" reigned in Germany.[66]

None of this furthered German foreign policy. As I have noted, Zimmermann's proposal generally met with support from right-wing, nationalist circles within Germany. Germany's supporters abroad, however, were horrified. George Viereck, a prominent exponent of pro-German views in the United States, initially believed Zimmermann's dispatch was a hoax: "It is impossible to believe that the German Foreign Secretary would place his name under such a preposterous document . . . If Germany were plotting against us she would hardly adopt so clumsy a method. The *Realpolitiker* of the Wilhelmstrasse would never offer an alliance based on such ludicrous propositions as the conquest by Mexico of American territory."[67] Zimmermann's admission of the telegram's authenticity dashed such faith in the German cause.

Zimmermann's proposal to Mexico depended upon secrecy to avoid arousing international hostility. Such policies became less realistic in a more transparent world. Recent German history predisposed German leaders to take ruthless actions. One of Bismarck's unfavorable legacies to Germany's political elite was the sense that extreme national egoism and immorality constituted the height of statesmanship and realism. To a considerable degree, this perception resulted from a misreading of Bismarck's foreign policy, which was marked by both a sense of proportion and the maintenance of good relations with other countries.[68] Zimmermann's telegram demonstrated the fallaciousness of the supposed association between immorality and realism. A better-informed observer would have realized that Zimmermann's proposal was unlikely to produce any advantages for Germany and merely demonstrated that

German decisionmakers possessed scant understanding of events in the United States, Mexico, and Japan. Given the importance of justifying major foreign policy decisions to large masses of people, there was considerable realism in Bernstorff's statement during the Nazi era: "I am convinced that politics and morality are indissolubly wedded, and that a policy that is not guided by moral considerations will find no mercy before the tribunal of world history, though it may achieve a passing success."[69]

The first of the three observations, concerning the vulnerability of telegrams to interception, is perhaps the most obvious lesson of the Zimmermann incident. This episode also demonstrates the German government's responses to two other characteristics of the telegraph: its expense and its tendency to garble messages. In this instance, the German foreign ministry mishandled these two problem areas, although not with the disastrous consequences that resulted from its security failure. In regard to expense, the German embassy in Washington had orders to encrypt the telegram to Mexico into a second code, the concise, publicly available "A-B-C-Code." Like many organizations, the Auswärtiges Amt used commercial codes to diminish telegraph fees by reducing the size of messages. The embassy staff, however, neglected to employ this second code when sending the Zimmermann telegram to Mexico. Partly as a result of this oversight, the telegram, transmitted between Washington and Mexico City as a "Fast Day Message," cost a hefty $85.27 (about $1,148 in year 2000 dollars). In response to the problem of garbling, the German government had often sought to reduce the chance that a telegram would be delayed or scrambled by sending it to Washington via two or more routes. It did not do so with the Zimmermann telegram, and the German embassy in Washington struggled to decipher it.[70]

Disregarding the merits of the policy that underlay the Zimmermann incident—and analyzing it only as an example of diplomatic communication—it was a fiasco. The Zimmermann telegram reached Mexico City, but only at great expense, after having been garbled and laboriously reconstituted, and, worst of all, with its contents available to enemy espionage agencies. These failures are sobering, occurring as they did more than sixty years after the first use of the telegraph in diplomacy.

Technical and Economic Factors

7

Through trial and error, diplomatic users of the telegraph became acquainted with its characteristics. The telegraph challenged foreign policy establishments by greatly exacerbating three preexisting concerns: message security (intercepted dispatches), expense, and message integrity (garbling). These problems interacted with one another. As is frequently the case with technical-bureaucratic systems of some complexity, foreign ministries often discovered that their responses to one dilemma aggravated other difficulties. In general, the solutions to the problems of security and expense complemented each other but conflicted with the solutions to the problem of garbling.

Espionage

The use of the telegraph for diplomacy produced a new era in intelligence gathering. Whereas governments had once relied primarily upon spies, they increasingly found that signals intelligence was their best means of collecting secret information in an age of electric telecommunications. Diplomats responded to this shift by adopting codes to keep their telegrams secret. Codemaking was difficult and labor intensive, however, and foreign ministries struggled to conceive codes that would be cost-effective, concise, secretive, reliable, easy to use, flexible enough to cover any contingency, yet sufficiently different from preceding codes to erase the gains made by enemy codebreakers.

Previous to the telegraph, foreign ministries, while sometimes using codes, had also attempted to improve security through a more systematic use of special couriers, diplomatic pouches, and intricate wax seals. In September 1849 Alexis de Tocqueville, the French foreign minister, announced the establishment of a courier service that would provide "perfect security" between Paris and the French legations in Vienna and St. Petersburg.[1] Such tactics thwarted the black chambers (agencies that intercepted and decrypted messages) operating out of the governmental post offices of most European powers and made illicit perusal of the diplomatic mails more difficult.

Electric telegraphy reversed this trend toward greater message security. The secretary of the French Atlantic Telegraph Company aptly referred to telegrams as "open letters." Government management of the telegraph network facilitated espionage by state intelligence services. Great Britain nationalized its telegraph system in 1869, leaving the United States as the only major power with a privately owned system of electric communication.[2] Diplomats stationed in foreign countries were wise to assume that their host governments had access to any telegrams they received. Thomas F. Bayard, the U.S. secretary of state, warned, "It is generally understood that some governments, especially those whose telegraphs are under the charge of the central administration, spare no effort to decipher the secret messages of foreign representatives." A former Austrian code officer, describing the years before the First World War, noted, "Our telegraphers, like those of other governments, were instructed to send us a copy of every cipher dispatch that passed over their wires." During the First World War, the Vatican was the most striking example of a diplomatic power vulnerable to intercepted messages; once Italy entered the war, at least one of the belligerents, and frequently more, could obtain every telegram from the Papacy.[3] Under such circumstances, codes and ciphers were the only means of maintaining the confidentiality of messages.[4] Consequently, governments made significant advances in encryption and its twin, codebreaking. But some made greater advances than others.

Russia, France, and Austria-Hungary, in that order, led the world in communications intelligence before the First World War. During the reign of Nicholas II, the last tsar of Russia (1894–1917), the Russian government obtained decryptions of telegraphic dispatches from virtually all foreign ministries of any importance. Russian intelligence, among its most significant accomplishments, deciphered communications between Britain and Japan concerning the negotiations that preceded the alliance they concluded in 1902.[5] Russia also broke codes of the Austro-Hungarian foreign ministry. This allowed Russian statesmen to read secret Habsburg telegrams that belied that country's official policy in July 1914, probably strengthening Russia's willingness to take a hard line during the crisis that preceded the First World War.[6]

A secretary at the British embassy in St. Petersburg described the most basic technique used by Russian cryptanalysts:

> Whenever a Foreign Representative received a telegram in cipher from his Government, a copy of it was taken at the Ministry of the Interior. If the Representative receiving the telegram communicated to the Minister of Foreign Affairs anything a day or two later, this, too, was sent to the Ministry of the Interior. The communication and the telegram were then compared. Supposing the cipher to be a series of numbers, as is usual, a few words would thus be made out. The clues thus obtained would increase according to the frequency of the telegrams received, and in a very short time it would be possible to read the whole cipher.[7]

Spies trained to steal the ciphers of foreign embassies expedited the efforts of Russian codebreakers. In 1905 the Swedish and American governments learned that Russian agents had obtained their diplomatic codes. At about the same time, German officials had similar suspicions about Russian code burglary although they lacked conclusive proof. In 1906 the secretary of the British embassy, alarmed by the Russian proficiency at code stealing, took to sleeping near the safe containing the ciphers, and, as a further precaution, "A piece of thin paper is gummed every night across the entrance fastening the door."[8] Such concern with security produced excesses. One employee of the British embassy, after hurling himself at an apparent intruder in a darkened room, found that he had seized the ambassador by the throat. A British diplomat in St. Petersburg described British security measures as wasted effort, "given the highly developed talent of the Russian *cabinet noir* for finding out cyphers." Despite energetic countermeasures, Russian intelligence continued to intercept the telegrams of Britain and other nations. In 1916 a Russian official expressed concern that Britain, then a wartime ally, used ciphers that were read as easily as a "newspaper."[9]

This knowledge of foreign diplomatic codes frequently embarrassed Russian statesmen. During the mid-nineteenth century, Prince Gorchakov, the Russian foreign minister, after conveying information to the British ambassador, indicated that he wanted this information sent to the British Foreign Office by messenger rather than by telegraph. Gorchakov hoped to keep the information secret from his own country's Ministry of the Interior, which was reading British telegrams. Likewise, in 1899 Count Sergei Witte, the Russian finance minister, advised the German ambassador to Russia to be careful with his letters, whether ciphered or not. Another such incident occurred on 2 June 1904, when a prominent Russian politician gave the British ambassador to St. Petersburg "a disagreeable shock." He warned the diplomat that Russian intelligence read "all" of the British telegrams, and that he should send reports of their conversations by diplomatic pouch instead.[10]

The French Cabinet Noir also earned a reputation for its achievements in signals intelligence. It had broken telegraphic codes used by Italy, Britain, the Ottoman Empire, Germany, Spain, and Japan during the years before the First World War. Decryptions of secret telegrams sometimes shaped the course of French diplomacy. In October 1913 Italy approached France and Britain with a proposal for a Mediterranean accord. The two Entente powers hoped that this proposition evidenced a growing rift between Italy and Germany. The French government, however, indignantly broke off these conversations when it learned, by intercepting and decoding Italian telegrams, that Italy had negotiated in bad faith and reported the details of the talks to Germany.[11]

Unfortunately, the political instability and self-seeking atmosphere of the French Third Republic produced a poor environment for secret intelligence. Théophile Delcassé, the French foreign minister during the First Moroccan Crisis, is a noteworthy exception to this generalization. Delcassé showed great integrity by resigning in 1905 rather than keeping himself in power by releasing confidential information gleaned from German telegrams.[12] In contrast, opponents of Joseph Caillaux, the French prime minister in 1911, used decryptions to undermine his efforts to reach an agreement with Germany during the Second Moroccan Crisis. Caillaux was vulnerable to the decryptions because they revealed that he had gone behind the back of his foreign ministry and the French public in offering secret concessions to Germany. While defending himself during this controversy, Caillaux alerted the German embassy about the broken code and intercepted telegrams. As a result of these maneuverings, the German foreign ministry recognized the weakness of its communications security. The subsequent improvement of German codes left France unable to read German messages on the eve of the First World War. To make matters worse, France's ally Russia was also unable to break the new German codes.[13]

As is often the case with secret intelligence, the information gathered by the French Cabinet Noir sometimes obscured more than it illuminated. French codebreaking before the First World War relied upon the precarious genius of Étienne Bazeries, a poorly educated alcoholic whose brilliant intuition sometimes missed the mark. Compounding possible errors, French statesmen of the time romanticized the activities of codebreakers and tended to accept their work uncritically. The director of political affairs at the French foreign ministry on the Quai d'Orsay described the Cabinet Noir as a mystical place "where perspicacity, clairvoyance, penetration, flair, the strange instinct which lays bare enigmas and hieroglyphics, border on magical divination." France's ambassador to Italy complained: "Our entire policy is based on the [decryptions]. The reports of our ambassadors count for nothing." The Cabinet Noir inadvertently misled French officials by presenting them with intercepted telegrams wrenched from the broader context that made them comprehensible. For example, French intelligence decrypted a telegram in the autumn of 1887 that seemed to reveal Britain's involvement in duplicitous dealings with Germany behind the back of France. But if one knew the course of the negotiations, it was clear that the telegram actually attested to the reverse: British suspicion of Bismarck.[14]

In addition to providing an inadequate context, the Cabinet Noir sometimes passed on erroneous information. In September 1898 Delcassé, the French foreign minister, became alarmed after the Cabinet Noir gave him an Italian telegram incorrectly stating that the British ambassador was about to present him with an ultimatum addressing a dispute over the upper Nile. When he did not receive the ultimatum, Delcassé mistakenly believed that he had achieved a great diplomatic success for France. In another instance, the Cabinet Noir intercepted and partially deciphered an Italian telegram that seemed to contain evidence of the guilt of Captain Alfred Dreyfus, a French soldier of Alsatian Jewish origins unjustly accused of treason. Army officers convinced of Dreyfus's guilt circulated this incriminating version, and then suppressed a later, correct rendering that implied his innocence. The fallacious decryption was difficult for Dreyfus's defenders to challenge because it could not be openly examined without threatening France's national security. Octave Homberg, who later directed the French foreign ministry's Cabinet Noir, argued that the Dreyfus affair demonstrated the importance of transmitting only those decryptions "of which I could irrefutably prove the accuracy."[15]

In 1914 the reputation of Austria-Hungary's black chamber owed much to past greatness. During the eighteenth and early nineteenth centuries the main Austrian intelligence service, the Geheime Kabinets-Kanzlei, possessed a system of informants, mail burglars, translators, and codebreakers that won notoriety as the best in Europe. Lord Palmerston, while British foreign secretary, filled his letters to Vienna with propaganda and insults intended for the Habsburg officials who, he believed, covertly intercepted and read his missives. But Austrian liberals saw the black chamber as a repressive bulwark of the old regime. With the revolutions of 1848, Prince Metternich, the Habsburg politician who acted as the patron and protector of the black chamber, fled into exile. The Geheime Kabinets-Kanzlei closed up shop shortly thereafter, an ironic development given the possibilities for espionage presented by the newfangled electric telegraph. The Habsburg regime reestablished a black chamber—which read enciphered French correspondence during the Crimean War—but it seems never to have regained its former preeminence in codebreaking.[16]

Despite this setback, Austria-Hungary did engage in clever codebreaking stratagems. After making some initial progress against an Italian code, Habsburg cryptanalysts found themselves stymied. So they planted a carefully phrased article in a Constantinople newspaper that they knew would be likely to attract the attention of the Italian military attaché in that city. The attaché encoded it verbatim and sent it by telegraph to Rome via Austro-Hungarian wires. The Habsburg codebreakers intercepted this message and expanded their knowledge of the

Italian code vocabulary. One Austro-Hungarian official recounted that his government used this subterfuge numerous times until it recognized more than two thousand words, "could read the Ambassador's dispatches with ease, and knew whenever he got a reprimand."[17]

Neither Germany nor Britain possessed agencies to collect diplomatic signals intelligence before the First World War.[18] Many German officials, extrapolating from past experience, believed that firepower and speed of military mobilization would be far more important to their nation's future. They held this view despite evidence that France and Russia intercepted and read their diplomatic messages. Chancellor Otto von Bismarck was not fazed by foreign espionage. With typical resourcefulness, he even sought to exploit it in his diplomacy. In 1887 Bismarck explained that he used ciphered telegrams as an indirect means of communication with the Russian government. Truly secret messages he sent by courier. Bismarck's successors continued this policy. German leaders found other clever ways to preserve the secrecy of messages. During the Second Moroccan Crisis, the German government circumvented French signals intelligence by sending its agent in Agadir his secret instructions buried in seemingly innocuous telegrams.[19]

German diplomatic code security was poor, even during the First World War. After one incident in which it appeared that Britain had read a secret message, Emil Propp, who worked for the German cipher bureau, asserted that a cipher was safe unless "it has fallen into the hands of the enemy." A particularly egregious security violation is described by a secret U.S. report on German diplomatic codes. The German foreign ministry transmitted a wartime message to Washington in one of its most widely used codes (13040). The German ambassador, Count Bernstorff, then handed the message, virtually unchanged, to the U.S. secretary of state. The formerly classified report declares, "There was no American cryptographic bureau in existence at the time when that message had been sent. If there had been, the usefulness of 13040 would have vanished from that moment."[20]

Britain had abolished its "Decyphering Branch" in 1844, following public outrage over the opening of mail from the exiled Italian nationalist Giuseppe Mazzini. During the early twentieth century Paul and Jules Cambon, France's ambassadors to London and Berlin, trusted the British government more than their own foreign ministry: both asked the British Foreign Office to deliver their private correspondence. Likewise, American diplomats in Russia placed their faith in English couriers, preferring them to the Russian mails. This faith seems to have been well placed, as the Foreign Office apparently made no effort to intercept the diplomatic telegrams of other countries. Nonetheless, British colonial authorities (especially the administration of India) and military services did explore signals intelligence during the first years of the twentieth century. This experience proved useful during the First World War. After the war, Britain's codebreaking ability, like those of France and Russia before the war, yielded tools for use in cabinet infighting and bureaucratic warfare within the British government.[21]

British message security did not present a formidable obstacle to foreign espionage. One Foreign Office employee recalled "how a foreign Sovereign had quoted to a British Ambassador from a Foreign Office despatch which he was by no means supposed to have seen." In response to information that an Ottoman agent had probably stolen a British cipher, the British foreign secretary, Lord Granville, played a prank on Turkish signals intelligence. "I have amused myself by concocting a message [to our ambassador at Constantinople] in this cypher—Telling him in a message in another cypher that he need not mind, if he cannot decypher it. The Sultan if he succeeds in reading it, will not like it."[22]

Until the First World War, the United States, like Britain and Germany, lacked an establishment to break diplomatic codes.[23] It was often the case that when a nation lagged in codebreaking, its own communications security was weak. This generalization certainly applied to the United States. American legations and consulates sometimes even lacked means of encrypting and decrypting messages.[24] On other occasions, important messages went out in plaintext. In 1872 someone from a telegraph office leaked uncoded American telegrams to the newspapers, disrupting sensitive negotiations with Britain. In 1915 an American diplomat in Constantinople complained that the Ottoman minister of the interior enjoyed making "jocular" references to "more or less confidential" telegrams that the State Department had sent in plaintext. Paradoxically, Thomas Jefferson, before he became president, had devised a system of encipherment far more difficult to crack than the methods of secret communication used by the State Department in the nineteenth century. He did not, however, grasp the excellence of his own system, and his efforts fell into oblivion until their rediscovery in 1922. Jefferson, like many of the founding fathers of the United States, paid great attention to the security of diplomatic dispatches. Americans of a later generation would choose instead to emphasize thrift and tradition when designing codes.[25]

On 3 August 1866 John Bigelow, the U.S. minister to France, reported on the success of the recently laid Atlantic cable connecting America and Europe. He warned William Seward, the secretary of state, that the new technology called for new precautions:

> While joining in the chorus of congratulation with which this triumph of modern science is received by the civilized world I deem it my duty to suggest . . . that the State Department provide itself at once with a new cypher . . . It is not likely that it would suit the purpose of the Government to have its telegrams for this Legation read first by the French authorities, and yet you are well aware that nothing goes over a French telegraph wire, that is not transmitted to the Ministry of the Interior.

Bigelow believed that traitorous employees had purloined the State Department's code shortly before the Civil War, possibly passing it on to "the principal European governments."[26]

Seward dismissed this story. He doubted that either disloyal government employees or foreign intelligence services could have reproduced the "cypher of the Department which . . . is believed to be the most inscrutable ever invented." Accordingly, Seward proclaimed that no "offer [of a new code] has been or is likely to be entertained." His confidence in the code's invulnerability displays ignorance of one of the most basic rules of cryptography: any code will become insecure if used long enough in an unmodified form. The State Department's code had first seen use more than sixty years earlier: in 1803 James Monroe had employed it while serving as envoy to France. Strangely enough, its longevity may actually have made it slightly more secure. Cryptanalysts at Britain's Decyphering Branch had broken it decades earlier but then misplaced the solution, although they did manage to reconstruct much of their earlier work.[27]

On 23 November 1866 the State Department sent its first encrypted message via submarine cable. In the message Seward expressed displeasure at Napoleon III's decision to delay the withdrawal of French troops from Mexico. If French cryptanalysts had not already broken the State Department code, Seward facilitated their work by also giving his dispatch to the American press, thus violating another of the most basic rules of communications security: the prohibition against sending a message both in cipher and in clear. The State Department did not consider security a major priority at this time. The major U.S. legations had fallen out of the habit of enciphering their dispatches, even during the Civil War.[28] When Seward finally decided to adopt a new code, as we shall see, it was for reasons of frugality.

In 1876 the U.S. foreign service began using codes known by the color of the codebook. The Red Code served as the State Department's primary means of encryption from 1876 until 1899. During the late 1890s the State Department decided to prepare a new code. It acted partly out of knowledge that codes become less secure over time and partly because of suspicions that the Spanish government could read it. A report that Spain possessed the Red Code reached the State Department during the spring of 1898, shortly before the culmination of the crisis that resulted in the Spanish-American War. Despite this warning, the Department continued to use the code during the crisis and the war that followed.[29]

In 1899 the State Department adopted the Blue Code. Its secrecy did not long survive. In July 1905 the American ambassador to St. Petersburg reported, "I have discovered beyond a doubt that the Russian Government has in their possession our entire cable code." A Russian spy in the employ of the American embassy had made photographs of the codebook.[30] Two years later, in June 1907, the Blue codebook of the U.S. legation in Bucharest disappeared. Not long thereafter the apparent thieves attempted to peddle photographs of it to interested governments. The Japanese Ministry of Foreign Affairs alerted the United States about one such attempted sale. The U.S. chargé d'affaires in Vienna marveled at this amateurish crime, noting, "it is difficult to understand that any great foreign Power would expose itself to the danger of secretly buying a copy of our code, when they are probably already in possession of the main features of it through other means." The American system of message security, based on widely distributed and identical codebooks, magnified such losses. The U.S. ambassador to Germany warned, "A single theft . . . in any part of the world where we maintain a mission renders it possible to read all messages to or from the Department and any Embassy or Legation."[31]

Despite these security lapses, the Blue Code remained the State Department's primary means of encryption until 1910. As late as 1912, the State Department, hoping to avoid the cost of a new code, unsuccessfully sought to persuade the Departments of War and the Navy to adopt the Blue Code for secret communication among departments. During the early years of the First World War, President Woodrow Wilson and his advisor, Colonel Edward House, used the Blue Code for their private correspondence. Not until 1916 did the U.S. ambassador to Vienna learn that the Blue Code was not secure.[32] In 1917 an American diplomat in Romania protested its continued employment: "As this code has been in use eighteen years, as two copies were lost, and as it has been exposed by transmission [of] presidential speeches there is practical certainty [that] this code is not [a] safe instrument for highly confidential communications." The Swedish government received a copy of the U.S. Blue Code for safekeeping when the Ottoman Empire and the United States severed relations in 1917. A State Department official noted that Sweden's knowledge of this code "is worth remembering." But it was not remembered, and the United States continued to use the Blue Code in messages to Stockholm.[33]

In 1910 the United States sought to improve security by adopting the Green Code. The designers of the new code, believing that "many, if not all," foreign countries were familiar with the Blue Code, suggested that the new code follow a different format so that it could not be reconstructed through knowledge of its predecessor. The State Department asserted in 1916 that it had "observed no evidence that the GREEN CIPHER is not a safe means of secret communication."[34] By 1917, however, some American diplomats suspected that Britain, Romania, and perhaps other countries, most notably Germany, possessed the American Green Code. After the United States entered the First World War, the former chief of the intelligence division of the Danish army's general staff reported his belief that Germany was reading telegrams dispatched in the Green Code by the U.S. legation in Copenhagen.[35]

The mind-set of those who implemented the State Department's communication security is vividly illustrated by fanciful stories told by one of America's great literary figures. James Thurber, a State Department code clerk during the First World War, declared that the Department's codes "were intended to save words and cut telegraph costs, not to fool anybody." The legends the code clerks circulated among themselves testify to their "cat's out of the bag" attitude and their tendency to doubt the value of their tedious labor. According to an apocryphal tale retold by Thurber, the Germans taunted Washington "about our childish ciphers, . . . suggesting on one occasion that our clumsy device of combining two codes, in a desperate effort at deception, would have been a little harder if we had used two other codes, which they named." According to another story, popular among the clerks, "six of our code books were missing and . . . a seventh, neatly wrapped, firmly tied, and accompanied by a courteous note, had been returned to one or another of our embassies by the Japanese, either because they had finished with it or because they already had one." This last story has perhaps some foundation in fact as German authorities had graciously returned a State Department codebook lost by the consulate at Leipzig.[36]

America was not alone in having problems with communication security; other governments that used cables or wireless telegraphy to send secret messages made some of the same sorts of errors. This was true of far more professional organizations, such as Britain's Royal Navy. The British Admiralty repeatedly emphasized, though often without effect, the need to avoid practices that compromised codes, such as partially enciphering messages, repeating the same message in different codes, or restating a message in cipher that had been sent in clear text.[37] Such sloppy practices greatly facilitated the work of enemy espionage; as the U.S. Navy later noted, "A poor code in the hands of a good coding officer is far safer than the best of codes in the hands of a poor coding officer."[38]

Like the foreign ministries of other nations, the U.S. State Department struggled to respond to the insecurity of telegraphic communication. Only with reluctance did diplomatic bureaucracies accept the possibility that their telegrams were read by foreign governments. Jules Cambon, France's ambassador to Germany, chided his government for believing in the secrecy of coded telegrams. When, he wondered, would the French foreign ministry cease to criticize foreign statesmen "in ciphered telegrams that are surely deciphered? This is bad diplomatic manners: . . . first of all, it is childish, and then it risks injuries of pride which make it difficult, if not impossible, to resolve any questions."

Other countries would have benefited from such warnings. During the First World War, British naval intelligence intercepted a telegram from the German minister in Buenos Aires that described the acting Argentine foreign minister as "a notorious ass." Another telegram suggested that if German U-boats decided to attack Argentine steamers, they should sink them "without a trace" in order to avoid a public outcry. The publication of these telegrams increased anti-German sentiment throughout Latin America.[39]

Yet the influence of the telegraph upon signals intelligence also raises questions about the value of diplomatic secrecy. As the foregoing analysis suggests, sensitive international negotiations generally benefit from confidentiality. Bargaining becomes more difficult when the other side knows one's intentions. Moreover, the glaring scrutiny of excitable media and publics can create an unfavorable climate for politically sensitive discussions. Nonetheless, intercepted diplomatic telegrams seem to have had very slight influence upon international relations during the era before the First World War. As a seasoned Foreign Office clerk noted, "I cannot remember that, at any rate before the War, anything very serious happened in consequence of our ciphers becoming known."[40] Germany, the United States, and Britain lagged in the collection of signals intelligence and in the maintenance of message security without suffering devastating repercussions during the pre-1914 period. This is worth considering when weighing the disadvantages of secrecy—especially impaired decisionmaking and the abuse of power—against its advantages.[41] The leadership of Russia, France, and Austria-Hungary in telegraphic signals intelligence suggests that achievement in this area may have been a response to relative weakness and an outgrowth of domestic repression.[42]

Expense

Before the electric telegraph, secrecy was the principal purpose of cryptography, the art of writing messages in code or cipher. Telegraphy generated a new use for codes: reducing transmission costs. The second half of the nineteenth century saw the development of commercial codebooks designed in large part to help businesses lower their telegraph expenditures.[43] In the same way, encryption provided a means for foreign ministries to reduce costs. John H. Haswell, who designed several State

Department codes, noted: "Formerly, ciphers were used only in the transmission of secret communications and secrecy was the only element considered. Since the introduction of telegraphy, . . . expense, another motive, has been introduced and must be taken into account."[44] The high price of telegraphy probably had less effect upon the course of international affairs than did communication security. Nonetheless, telegraph costs had great impact on the day-to-day life of diplomats.

In addition to popularizing codes, expense shaped the form of telegraphic messages. The Western Union Company later noted, "The keynote of the well-worded telegram is conciseness." Not only were lengthy dispatches expensive, they also undermined the speed advantage of the telegraph, since telegraphy carried missives one letter at a time. In May 1848, several years after the construction of Morse's first telegraph line, a magazine article predicted: "The certain effect of the Telegraph . . . will be to introduce a style of writing which shall be, *first of all, brief* . . . The Telegraphic style, as we shall denominate it, . . . is also terse, condensed, expressive, sparing of expletives and utterly ignorant of synonyms."[45] Scholars have noted that telegraphy contributed to making journalism more succinct. Such famous writers as Ernest Hemingway and William Shirer have described how the style they learned as journalists affected their other literary endeavors. Shirer noted, "To write cablese you had to strip narrative of everything extraneous. [It became] the language I wrote in for many years. And while it apparently was good for Hemingway, it worked badly for me. It made for a stilted style, a barrenness of language. Pretty soon it was the only way you could write."[46]

The same pressure toward pithiness affected diplomats. Relative to European foreign ministries, the U.S. State Department paid particular attention to cost constraints. American diplomats eliminated articles, prepositions, pronouns, and sometimes even verbs from their telegrams. For example, in 1881, following the assassination of Tsar Alexander II, the U.S. minister to Russia sent the State Department a telegram that read simply, "Emperor dead."[47] Although the European powers were less affected by such concerns, considerations of expense also influenced the content of their messages. During the First World War, the secretary of the German legation in Teheran was asked "in the future to avoid every manner of longwindedness in his telegraphic reports, particularly all superfluous forms of courtesy." Similarly, at the British Foreign Office, the permanent undersecretary "took a malicious satisfaction" in returning the telegrams of diplomats "in an abridged form." Such demands damaged the literary quality of the best diplomatic reporting. The head of chancery at the Austro-Hungarian foreign ministry mourned the decline in "the art of style" that resulted from "the succinct brevity necessary in telegraphic intercourse."[48] The new emphasis on concision challenged the culture of diplomacy, which had been shaped by flowery language and aristocratic notions of courtesy. Some of these traditions served a purpose. Honorific titles and other forms of personal flattery created a hospitable climate for friendly relations. Ambiguity, imprecision, and long-winded indirect language made compromises easier when negotiating parties had substantial disagreements.

At times, cost concerns encouraged the elimination of useful information, and thereby introduced new ambiguities into telegrams.[49] One British diplomat noted, "Telegraphic instructions are very concise, and it may be difficult to understand them." In 1903 the *Kölnische Zeitung,* a German newspaper, complained that Germany's cost-consciousness had hindered its diplomacy during the second Venezuelan crisis, a dispute with the United States that arose when Germany and Britain sought to collect unpaid debts from Venezuela. The newspaper asserted that the brevity of German diplomatic telegrams left the kaiser's government unable to counter the propaganda emanating from the jingo press of the United States.[50]

Underwater (submarine) telegraph cables generally charged much higher tariffs than landlines. This was due to a number of factors: the enormous capital outlays necessary to build and lay a wire capable of withstanding the ocean environment, the difficulty of maintenance, and the dearth of competition that might have restrained acquisitive cable companies. Users faced the highest costs during the early days of the cable. In November 1851 a cable between Britain and France began operation. Count Alexandre Walewski, the French ambassador to Britain, had mixed feelings about this new method of communication. Initially enthusiastic, he asked his superiors to authorize the use of the telegraph for conveying urgent news between his embassy and the Quai d'Orsay. But after using the cable for a while Walewski complained of "ruinous" tariffs charged by the telegraph company. The exorbitant cost resulted in part from the lucrative monopoly granted to the Brett brothers, who laid the Anglo-French cable. Yet, despite the expense involved, Walewski continued to insist upon "the importance of frequent communication" by telegraph between himself and the French foreign ministry: "This importance is more and more felt every day, especially since big business, the press, and even the [British] Foreign Office have come to use this method of communication almost daily." During the Crimean War the British and French governments made extensive use of the cable. In 1855 the British embassy in Paris spent 20,214.44 francs (£809.0.4) on official telegrams, a very large sum that constituted 20 percent of the total embassy expenses reimbursed by the Foreign Office.[51]

After Walewski became foreign minister of France a few years later, he issued a circular to French diplomats blaming their excessive use of the telegraph for much of the alarming rise in the ministry's budget during the 1850s. He asked that they refrain from employing this mode of communication except in "exceptional and urgent cases." Similar circulars, disseminated in later years, indicate the persistence of the Quai d'Orsay's displeasure over long, inessential telegrams. Decades later the French politician Raymond Poincaré instituted a policy of making diplomats pay for telegrams considered excessive in length or frequency.[52]

It is ironic that in 1858 the British Foreign Office expressed the hope that "the use of the telegraph . . . , if judiciously employed, ought materially to lessen the expense incurred on account of Messengers." The heavy expenses of the queen's diplomatic couriers had aroused concern. Yet the idea that the telegraph would serve as a cost-cutting technology presupposed not only the elimination of "useless words" but also that "telegrams should be used only to ask questions, and to give information, directions, short explanations, and acknowledgments." The large sums spent on telegrams (£5,912 in 1857) already indicated the difficulty of implementing this strategy of sending only concise messages.[53]

As with the Anglo-French cable, diplomats using the Atlantic cable initially found, in the words of one British bureaucrat, "that the present charges are prohibitive." Customers initially faced a tariff of $10 a word (worth more than $100 per word in the year 2000, when adjusted for inflation).[54] During the early years, cable companies faced little price competition on the transatlantic route. The Anglo-American Telegraph Company owned the first two successful cables, both completed in 1866. In 1869 a French enterprise, the Société du Câble Transatlantique

Français, laid a third cable. The two companies avoided a price war by forming a cartel. Elihu Washburne, the American minister in Paris, complained in 1875 that the transatlantic cable "monopolies," unfettered by competition, had engaged in a campaign of "robbery and extortion" against the governments and people of America and Europe.[55]

The Anglo-American Telegraph Company levied particularly "exorbitant rates" for messages in code. The company charged coded telegrams at twice the rate of regular messages. It defended this policy by noting that coded dispatches, prone as they were to garbling, needed to be repeated back for accuracy. To make matters worse, Britain and the United States, like many foreign ministries, used codes that converted plaintext into numbers. The telegraph company penalized numerical codes by expressing all numbers as words and charging accordingly. The British minister in Washington remarked: "It follows, therefore, that as, according to the systems of cypher in my possession, each word has in most cases to be represented by four figures, the cost of sending it across the Atlantic is multiplied eight times. When besides this, words have to be spelt, as is the case with most names, there being no exact equivalent for them in the cyphers, the cost is indefinitely increased."[56] The State Department code produced even greater expenses than that of the Foreign Office; compared with an uncoded message, it multiplied the number of counted words about fivefold and the expense (because of the repetition of all coded messages) tenfold.

The issue of expense carried particular resonance in the United States, where the republican tradition produced an abhorrence of public expenditure. For the State Department, as one historian has written, "international comity and efficiency were always secondary to considerations of economy."[57] Motivated mostly, it seems, by cost considerations, the Department had explicitly rejected diplomatic couriers "such as are regularly attached to the Departments of Foreign Affairs of European Governments." Likewise, the State Department sought to reduce postage costs by discouraging the unnecessary use of wax seals. During the U.S. Civil War, the secretary of state, seeking means to fund the Northern war effort, advised American diplomats "to use HALF sheets of paper in all cases where they will suffice to contain the text of the note to be copied. This will save unnecessary waste of paper, and largely diminish foreign postages . . . Consuls, whose forms and returns are frequently brief, may with propriety cut off that portion of the paper which remains blank."[58]

This frugality carried over into the State Department's attitude toward the telegraph. American diplomats dreaded the appearance of financial extravagance. In July 1870 Hamilton Fish, the American secretary of state, only learned of the developing crisis between France and Prussia when foreign diplomats asked him about the U.S. attitude toward it. When news of the French declaration of war reached the American government, it came via the Associated Press. The U.S. minister to France in 1870, Elihu B. Washburne, had presidential ambitions and hoped to avoid creating the impression that he squandered public funds on telegrams. His legation sent time-sensitive dispatches to the secretary of state by steamer; their warnings of the dispute over the Hohenzollern candidature for the Spanish throne and then of the complications posed by the Ems telegram both arrived after the French declaration of war.[59]

The instructions issued to American diplomats in 1885 and 1896 permitted the use of the telegraph only in cases of emergency. These rules required that the head of mission at his own expense cover the cost of telegrams that the State Department deemed inessential. American diplomats did receive leeway to make full use of the telegraph during the Spanish-American War.[60] Nonetheless, the long-term emphasis on parsimony continued.

On 4 June 1914 William Jennings Bryan, the U.S. secretary of state, issued an order to employees of the State Department: "The use of the cable is restricted to the transmission of messages of great importance and urgency, or to messages despatched under instructions from the Department of State. All such messages should be as briefly and concisely written as is possible to make them intelligible. The use of the cable at the expense of the Government in the ordinary business of the Department . . . is strictly prohibited." One month later, on 13 July, Frank Mallet, U.S. vice-consul general in Budapest, saw many harbingers of the upcoming war between Austria-Hungary and Serbia; his prescience set him apart from other American diplomats, who did not perceive the impending crisis. At this point Mallet faced a dilemma. His superiors would be likely to rebuke him for sending such a long message—about 150 words—by telegraph. The State Department considered long telegrams financially extravagant and an excessive strain upon those charged with deciphering them. Accordingly, Mallet sent a letter rather than a telegram. His warning did not arrive until fourteen days later, on 27 July. Austria-Hungary declared war on Serbia the next morning.[61]

The First World War undermined the culture of frugality that had prevailed among U.S. diplomats. Following the war, however, when it "no longer [had] the unlimited funds available during the war," the State Department complained that "diplomatic and consular officers have fallen into the habit of writing telegrams, where under normal circumstances the mail would suffice." It also noted the decline of concision: "The habit of writing telegraphic messages in epistolary form . . . prevails to a great extent." Despite such pleas, American diplomats' use of the telegraph never returned to its former, relatively modest level. Like its European counterparts, the State Department had accepted telegraphy as an everyday means of communication.[62]

The State Department adopted telegraphy more slowly than the foreign ministries of other great powers, but all of them worried about the cost of electric telecommunications and the monopolization of the wires by unimportant messages. They especially sought to monitor and control messages carried over intercontinental submarine cables. In the 1880s the British Foreign Office found the cost of cablegrams between London and the Pacific coast of South America unsustainable: nine shillings (more than two dollars) per word. Likewise, the Foreign Office saved money by discouraging its officials from using the telegraph to communicate between London and the Far East.[63] During the budget controversies of the early twentieth century, political authorities in imperial Germany tried to halt the alarming rise in the telegraph expenditures of German diplomats. The secretary of the Auswärtiges Amt decried the sending of verbose telegrams that were not urgent.[64] In 1903 the Budget Commission of the Reichstag encouraged thrift by cutting the appropriation for the Auswärtiges Amt's telegraph expenditures by 15,000 Marks (about $70,000 in the year 2000). Such concerns became more acute during the First World War; the German chancellor's office threatened to punish and exact compensation from diplomats who sent "useless and unnecessary" telegrams at a time when the "utmost economy is demanded."[65]

Time and experience ameliorated some of the problems posed by expense. Telegraph tariffs declined as the technology continued to improve, competition increased, and companies sought to attract new customers. Notwithstanding the deflation of the late nineteenth century, the decline in prices for Atlantic cablegrams was striking. The cost of using the Atlantic cable dropped at an average rate of 15 percent per year during the first twenty years of its existence, from about $10 per word in 1866 to 40 cents in 1886. By 1904 it was 20 cents. The decline of tariffs between Europe and East Asia was much less pronounced (about a 50 percent reduction, in an era of general deflation) because of a lack of competition on the far eastern routes.[66]

Foreign ministries gradually adapted to the high cost of telegraphy through privileged arrangements with telegraph companies, concise codes, and regulations about the format and language of telegrams. Ministries using both the Anglo-French and the Anglo-American cables successfully bargained with the cable companies for cost reductions. The German foreign ministry considered sending some messages by a combination of telegraph and post to achieve an optimal balance of speed and frugality.[67] The U.S. State Department arranged to convey many of its telegrams during off-peak hours at a lower rate, and prescribed shortened forms of expression. The State Department also discovered that Washington-to-Paris was the cheapest route for sending telegrams across the Atlantic. It exploited this fact by using Paris as a distribution center for telegrams to many European and African cities.[68] Such practices eased, although they certainly did not eliminate, the financial burden associated with diplomatic telegrams.[69]

GARBLING

Let us return once again to William Seward and the 1866 Atlantic cable, this time to illustrate how security and cost concerns influenced a third major challenge associated with diplomatic telegraphy, that of garbled telegrams. As we have seen, the pseudo-secrecy of the State Department code used in 1866 did not prompt its displacement by a more secure method of encryption. The code's tendency to dramatically increase the cost of messages, however, captured the attention of Secretary of State Seward. The State Department had historically distinguished itself by its parsimony, but cost-cutting was all the more necessary because of the huge debt incurred by the U.S. government during the Civil War. The

State Department's overseas budget shrank from $140,000 for the fiscal year ending in June 1866 to $115,000 for 1867.[70]

As discussed earlier, Seward sent the State Department's first encrypted telegram on 23 November 1866. Transatlantic telegrams cost "only" about $2.50 a word at this time, after a 50 percent cost reduction on the first of the month. The coding, however, enlarged Seward's dispatch from 780 words in plaintext to 3,772 words as counted by the telegraph company. On top of this, the company charged double for messages in code. To his astonishment, Seward learned that the cablegram to France had cost $19,540.50, more than three times his own yearly salary. He refused to pay this enormous sum, and the telegraph company had to win a lawsuit against the State Department before it finally collected its toll, five years later.[71]

In response to this debacle, Seward "set to work as early and prosecuted as vigorously as possible the construction of a new and frugal cipher code." These efforts yielded a more concise code, which the Department used from 1867 until 1876. The new code shortened messages by substituting short letter groups—containing from one to four letters—for words. Seward expressed delight over this code, which, he noted, communicated "every word in the English language . . . in an average of less than three letters." In practice, the code proved concise and provided great security against codebreakers.[72]

Unfortunately, while the new code reduced costs (by reducing message length) and increased secrecy, the State Department had not considered the needs of telegraphers and code clerks. As a result, the State Department's new code of 1867 defied not only foreign cryptanalysts but also the American diplomats who used it. Many of the letter groups had double meanings: "b-a-g," for example, meant both "word" and "negotiation." The lack of an alphabetical listing of the letter groups made decoding very time consuming. Most important, the code words did not contain a standard number of letters and became very difficult to decode if telegraphers ran the groups together, as they often did in the pursuit of speed.[73] One State Department clerk wrote to Hamilton Fish, Seward's successor as secretary of state: "It will not perhaps have escaped your recollection, that the first cipher message as received at the Department from our minister to Turkey formed one long string of connected letters, which for a time was considered by many in the Department as a conundrum, but finally, after considerable labor was deciphered . . . A telegram was received in a similar condition . . . from Vienna, the latter I believe has never been deciphered." Scrambled dispatches also plagued communication with the American ministers to Russia, France, Spain, and Britain. After a jumbled message from London, Fish remarked upon "our sadly incomprehensible cipher."[74]

While the cable was only to be used for urgent messages, the State Department sometimes could not decode the garbled telegrams until after the same messages arrived by post. Likewise, garbling delayed negotiations as diplomats waited for "clearly written instructions" to arrive in the mail. Robert Schenck, the U.S. minister to London, was so frustrated "by having to correspond through the difficult and unsatisfactory medium of the telegraph" that he hoped to rid himself of responsibility for one topic of Anglo-American negotiation by having its site transferred from London to Washington.[75] Seward's pursuit of lower cable costs had produced false economies.

Seward complained to the Anglo-American Telegraph Company about the "frequent and in some instances important" errors that, he charged, had resulted from their negligence. Seward wrongly asserted that such instances "cannot be ascribed to any complication in the cipher itself." The secretary of state's response was typical. In cases of garbled telegrams, as with breaches of communication security, foreign ministries tended to blame incompetent or disloyal clerks rather than considering the vulnerabilities of their codes or of the technology itself. During the First and Second World Wars, Germany made the same mistake, pursuing traitorous individuals rather than reevaluating its communication system, after British signals intelligence read German messages in such famous cases as the Zimmermann telegram.[76]

The State Department's 1867 code provides an extreme case that illustrates a general fact: use of the telegraph greatly increased the risk of garbled messages.[77] During the early history of electric telegraphy, users came to expect errors. One historian who examined European diplomacy during the early 1850s declared, "The telegraph worked both uncertainly and inaccurately, and was seldom trusted until a dispatch arrived." An article in the Washington *Daily Union* noted, "All the intelligence . . . now published is obtained through the medium of the magnetic telegraph; it is not to be expected, therefore, that every word . . . is literally correct." The use of the telegraph, one editor commented, "requires mighty good guessing." The *New York Herald,* after receiving an obviously fallacious report, remarked upon "the vagaries to which the telegraph is sometimes liable." Mistakes were sometimes costly; damage suits due to scrambled telegrams influenced the development of contract law during the nineteenth century.[78]

The causes of garbling were many and varied. The equipment malfunctioned at times, although technological improvements, such as better means of insulating the wire, considerably reduced these errors. Submarine cables produced more mistakes than landlines because the surrounding water blurred the electric impulse.[79] Atmospheric conditions—such as electrical storms and the northern lights—scrambled messages, at times completely disrupting communication. Other acts of nature could wreak havoc on telegraph lines. A snowstorm on 19 November 1909 damaged much of Germany's telegraph (and telephone) network, isolating Berlin from the rest of the world. According to the RCA Annual Report for 1930, "An earthquake in the bed of the ocean snapped 12 cables in the North Atlantic on November 18th, 1929." Bad weather in the Rhine valley on 9 July 1870 truncated, delayed, and distorted telegrams, hindering French diplomacy during the crisis that preceded the Franco-Prussian War.[80]

Errors were sometimes intentional. In 1915 a Romanian telegraph clerk, apparently on his own initiative, started garbling German telegrams because he thought war had broken out between Romania and the central powers. He was relieved of his position after a German diplomat complained. More often, faulty communication resulted from human error. Telegraphers sometimes made mistakes when sending, receiving, or transcribing messages. Morse code, the most widely used system for translating the electrical impulses into letters, placed heavy demands on operators and equipment. Small variations in the length of dots, dashes, and spaces reduced accuracy. One telegraph company found that two problems, the loss of a dot in transmission and false spacing, accounted for the vast majority of its errors.[81]

Illegible handwriting among operators and scribes also produced problems. The general staff of the German navy alleged "that most errors in the transmission and repetition of ciphered telegrams result from the various means of writing individual letters." In response, it attempted to standardize handwriting among clerks. An American diplomat in Nicaragua claimed that the "execrable" handwriting of telegraphers in his host country contributed to a constant stream of garbled cables. Charles Francis Adams, as U.S. minister to Britain, advised that when sending and receiving telegrams, "it will be of the greatest importance that the letters should be written in a very clear round hand, admitting of no confusion." In one of its codes the State Department purposely omitted the letter *I* because it resembled the letter *J*. At the German foreign ministry, Bismarck went to great lengths to improve the quality of handwriting during the age before the typewriter.[82]

Errors became more common when telegraphers handled messages that they found meaningless. Deprived of contextual clues, they could not rely on the sense of a message to help them interpret ambiguous characters. Operators had trouble with telegrams in foreign languages. A Brazilian diplomat complained that British telegraphers "almost always garbled" telegrams in Portuguese. The American Red codebook of 1876 remarked that telegraphers from China, France, Germany, Italy, Japan, Russia, Spain, and Turkey, "ignorant of each other's languages, . . . constantly commit vexatious and often serious mutilations of original messages." The State Department's 1867 code omitted the letter *W*, partly "because it is apt to be obscurely written, partly because it is unnecessary, but mostly because, as it is not a letter used in the languages of continental Europe of Latin origin, it would tend to puzzle and mislead the operators in that quarter."[83]

Telegraphers found messages in code even less comprehensible. The difficulty of accurately transmitting encrypted telegrams troubled the company that operated the first Anglo-French cable. In the early days of the Atlantic cable, the Anglo-American Telegraph Company had its operators repeat back all encrypted telegrams in an effort to improve accuracy. After the British government complained about the resultant doubled cost of encoded telegrams, the company agreed to forgo the extra charge on the condition that it "shall not be responsible for any error in messages not repeated." Benjamin Moran, the chargé d'affaires at the U.S. embassy in London, exclaimed after one incorrectly transmitted message, "The failure of the telegraph company in this case is such a serious blunder as to raise grave doubts as to the propriety of trusting it with cipher telegrams at all."[84]

Telegraph distribution room at the Paris Peace Conference, February 1919.
National Archives, photo no. 111-SC-41447

Foreign ministries gradually developed ways to minimize the risk of garbling. A French diplomat, whose career began after the First World War, later noted that the Quai d'Orsay had proscribed the use of the "telegraphic" style. German diplomats received instructions to send "non-secret telegrams *of great importance*" in cleartext rather than code.[85] The practice of repeating back coded messages, despite the extra expense involved, helped prevent errors. Western Union employed this practice for State Department cipher cablegrams, and a regulation of the 1875 International Telegraph Conference at St. Petersburg stated that ciphered messages must be repeated.[86] Recapitulation of crucial phrases or words reduced risk of misunderstanding without greatly increasing the length of telegrams. For example, because the accidental addition or subtraction of negations disproportionately affected the meaning of sentences, U.S. State Department telegrams during the Cold War repeated the word "not."

The State Department also considered making greater use of numbers, which were intelligible in any language. Foreign ministries sought to confirm the accuracy of telegrams by sending the same message by post, although this facilitated the work of foreign codebreakers. Despite such measures, in 1908 Elihu Root, the U.S. secretary of state, complained that "errors in the transmission of telegrams occur frequently." During the 1950s a recently retired British diplomat remarked, "The means of communication are still far from perfect, and are liable to give rise to complications and misunderstandings even when the greatest care is exercised. Particularly is this so of telegrams which are sent in cypher."[87]

The expense of the telegraph encouraged concision, reduced redundancy, and, ultimately, multiplied the consequences of garbled telegrams. In 1903 Bernhard von Bülow, the German chancellor, discoursed on the need to avoid prolixity without creating misunderstandings. This was easier said than done. The example of the State Department, which faced cost constraints of particular severity, illustrates the dilemma. An American diplomat entered into a dispute with the Department in 1878 over the policy of transmitting coded telegrams without punctuation. He complained that this policy impaired comprehensibility. The State Department replied that punctuation was expensive and, because of its predictability, made code breaking easier.[88]

The pressure to shorten diplomatic telegrams exacerbated the telegraph's tendency to interfere with the exact reproduction of messages. Painstaking precision was required of copy clerks in the age before the Xerox machine. Telegraphy threatened the integrity of dispatches by requiring that they be recopied numerous times, often by clerks with no knowledge of their meaning. The increased use of encryption produced additional opportunities for errors by clerks assigned the slow, boring, and sometimes difficult task of enciphering and deciphering dispatches. Conscious of such dangers, one U.S. secretary of state warned "that the Department expects the accuracy of all telegrams in cipher to be carefully verified before transmission."[89]

Errors sometimes had dramatic consequences. In 1884 Bismarck recognized the exactitude required of code clerks: "I believe that there is no single position in either the Prussian or Imperial [German] service, in which such an amount of work, such discretion, and such precision is demanded at all hours of the day from subordinate officials in exchange for

. . . moderate pay . . . A single mistake, an incorrect collation of a telegraphic dispatch exchanged between powers in a [delicate international environment], what a disaster that can cause!" Bismarck knew whereof he spoke. A misdeciphered telegram at the Prussian legation in Madrid was the immediate cause of the Franco-Prussian War. The mistake seemed slight: a substitution of the word "9th" for the word "26th." Yet this error postponed Bismarck's secret plot to place a relative of the Prussian king on the throne of Spain. In the interim, the French government learned of the scheme, thus initiating a crisis that led to the war. The tendency to attribute this accident to the error of a German clerk obscures the responsibility of those who designed and oversaw a system so lacking in redundancy that a trivial failure could trigger a major crisis.[90]

Unsent Telegrams

As we have seen, the risk of interception, expense, and garbling affected the form and the use of diplomatic dispatches. Some telegrams were never sent at all because of these concerns. It is difficult to reconstruct such instances, since one is generally looking for something that is noticeable by its absence. One can nonetheless find traces of unsent telegrams in the historical record. I have already discussed American diplomats who, obsessed by the cost of telegrams, failed to warn the State Department of the approach of war in 1870 and 1914. Half a year after the completion of the 1866 Atlantic cable, a French envoy remarked, "The price of communications by transatlantic cable are so high that I always hesitate . . . to utilize this form of correspondence." In 1907 a German diplomat in Mexico stated that he found himself continuously wondering whether a given event was important enough to justify the expense of using the Atlantic cable to report it. Likewise, in 1895 during the first Venezuelan crisis between Britain and the United States, the U.S. government repeatedly asked for a British reply to U.S. demands before President Grover Cleveland's annual message to Congress in December. The British government's response was too long (that is, expensive) to send by telegraph and too complicated to summarize. So it traveled by steamer rather than cable. As a result, although the U.S. embassy in London possessed the response well before the president's speech, it reached Washington several days after the deadline for its arrival.[91]

In a similar manner, anxiety about security deterred messages. The Duke of Gramont, France's foreign minister in 1870, told his ambassador to Berlin that some matters were "of a nature too sensitive to be discussed in a telegram," even one that was enciphered. According to that ambassador, the Prussian king used similar language when discussing the Hohenzollern candidature in July 1870: "He could not address such a delicate point over the telegraph." Likewise, during the First World War, the Vatican's nuncio in Germany feared to send highly confidential messages to the telegraph office because he suspected that German intelligence could solve his code. On the other side of the coin, security procedures themselves sometimes deterred the sending of messages. Lord Lyons, serving as the British minister to the United States, noted that "the labour and difficulty of using the cypher" resulted in dispatches that left out important information.[92]

The problem of garbling, too, occasionally dissuaded statesmen from using the telegraph. The botched enciphered telegram about the Hohenzollern candidature apparently so unnerved German diplomacy that the councilor of the Prussian embassy in Madrid sent an *uncoded* telegram to the candidate's father saying: "On account of difficulties of written and telegraphic communication His Majesty has sent a trusted army officer to Your Royal Highness for discussions by word of mouth." In another example, at the end of the First World War, Italian leaders were so concerned about the possibility of a garbled telegram eradicating hard-won territorial gains that they sent their armistice terms by courier rather than telegraph, thereby delaying the end of the war with Austria-Hungary by about two days.[93]

PRESTIGE

One other attribute of telegraphic communication, this one cultural rather than technical, deserves mention: prestige. Telegrams carried particular connotations because of such characteristics as expensiveness, rapidity, and association with scientific progress. Whatever else they said, telegrams conveyed a sense of wealth, power, and urgency. Consumers exploited the telegraph's snob appeal.[94] Telegraphy was a sure-fire way to show that one was modern. It embodied the achievements of science (much as computers would do for a later generation) and captured the public imagination. Newspapers capitalized on this attraction, heralding dispatches received by telegraph. Individuals could demonstrate their own importance by ignoring messages sent by regular post, on the assumption that busy people only had time to read telegrams. For example, this was the practice of the entourage surrounding Cecil Rhodes.[95]

Diplomats sometimes found that the only way they could get their superiors to read important but non-urgent messages was to send them by telegraph. At the other end of the wire, access to confidential telegrams were a perquisite for the powerful and an affirmation of status. The thrill of such messages was best communicated by those suddenly deprived of access to them. Rab Butler, who lost his position as British foreign secretary in the 1964 election, rued the loss of access to secret telegrams. Likewise, Winston Churchill, upon learning that he had lost the 1945 election, lamented, "No more boxes" (of telegrams). In the mid-twentieth century, a hundred years after its commercial adoption, the telegraph still carried a cultural resonance that impinged upon the world of diplomacy.[96]

Conclusion

What happened when diplomats, who tended to be conservative and wedded to tradition, encountered the telegraph, a potentially revolutionary mode of communication? Foreign ministries initially hesitated to adopt telegraphy. It seemed a bourgeois technology that subverted prevailing notions of decorum and threatened the identity of diplomats as gentlemen of leisure. Moreover, such factors as expense, the risk of interception or garbling, and institutional inertia made foreign policy bureaucracies wary of this new technology. In the long run, however, the adoption of telegraphy by diplomats contributed to a number of important trends: the increased bureaucratization and centralization of foreign ministries, the rising importance of signals intelligence, the declining autonomy of diplomatic envoys, and, perhaps most important, the accelerated speed of international crises. The faster pace of diplomatic disputes invited more emotional and less creative decisions on the part of statesmen, while public opinion, which sometimes moderates over the course of a long crisis, often exercised a belligerent influence on shorter crises. In addition to providing new capabilities, electric telegraphy placed new strains on foreign ministries and created new problems, the solutions to which sometimes produced difficulties of their own.

From the perspective of efficiency, diplomatic telegraphy brought many advantages. Much of the time, faster communication facilitated day-to-day diplomacy, allowing rapid answers to queries and timely solutions to sudden problems. Yet diplomats, from code clerks to ambassadors, did not always feel they themselves had benefited from the change. Indeed, many complained about a technology that reduced their authority, lessened their control over their work environment, and seemingly degraded their working conditions. Telegraphy, of course, was only one of many factors contributing to bureaucratization, the rising influence of public opinion, and pressures toward professionalization. Nonetheless, some diplomats fought against this tide, engaging in acts of resistance against the telegraph that occasionally had important repercussions.

Does the story of telegraphic diplomacy offer us useful lessons about the social effects of technology? Historians generally recoil from the view that events repeat themselves so exactly that knowing about the past enables one to divine the future.[1] Indeed, an examination of contemporary commentary about the telegraph indicates that clever, well-informed observers often made inaccurate predictions. Electric communication, like other technologies that provide dramatically new capabilities, generated effects that were complex and difficult to anticipate. Even problems that seem obvious in retrospect were frequently surprising to contemporaries. Some of the more conspicuous challenges associated with the telegraph—its insecurity, high cost, and tendency to garble messages—were alleviated mainly through trial and error. People generally responded to the unintended effects of telegraphy as they arose.[2]

Knowledge of past experience does not provide the gift of prophecy, but it can raise issues that improve the quality of our thinking about the social consequences of technological change. It can also alert us to information that will help us to understand such consequences. The story told in this book suggests four questions that we can ask when we consider revolutionary technologies. All of these questions involve the autonomy of individuals faced with profound, impersonal forces, and all provide analytical tools that are useful for thinking about any new technology.

First, is a new technology likely to change the way human beings conduct their activities? This question derives from the classic argument between "technological determinists," who believe that technology determines social, political, and cultural change, and "social constructivists," who believe that social, political, and cultural forces drive the development and use of technology.[3] The debate between these two viewpoints addresses fundamental questions of historical causation. To answer the question, one must consider the relationship between the capabilities of a technology and the way it is used in practice.

Three examples demonstrate that telegraphy standardized behavior much more in some areas than in others. First, different foreign ministries allowed their diplomats different degrees of autonomy. Technology influenced but did not determine policy in this instance. Second, diplomatic culture clashed with a seeming technological imperative—the tendency to do something faster when this is an option. Some diplomats displayed reluctance to adopt the telegraph and exploit its capability for speedy communication. Long-standing social and cultural forces exerted great influence over diplomats, reinforcing their refusal to let themselves be hurried. During crises, however, demands for speed became more insistent, and the pressure of events often forced diplomats and foreign ministries to adjust their behavior. Here one sees partial and sluggish adaptation to technological change. A third pattern is illustrated by diplomats' responses to such characteristics of the telegraph as its insecurity and cost. In this instance one sees the fairly rapid adjustment predicted by the model of technological determinism (for example, the use of codes). This adaptation, however, did not occur automatically, nor was it instantaneous.

These varying outcomes lead one to ask why telegraphy exerted more influence in some circumstances than in others. It wielded the greatest power in realms where pre-telegraphic behaviors (for example, verbose messages) had obvious and immediate adverse repercussions (such as high costs). Foreign policy bureaucracies, which tended to be conservative and risk averse, were strongly motivated to avoid negative outcomes. When not faced with a negative outcome, they tended to continue present practices and/or do nothing.

The second question is whether a technology tends to expand or restrict people's freedom to pursue their goals. Does a given technology act as a tool (extending human capacities) or a constraint? The proponents of a transatlantic cable anticipated that telegraphy would be an invaluable tool in diplomacy, allowing governments to settle in hours disputes or problems that might otherwise drag on for weeks or months. Improved media of communication can empower and increase the efficiency of individuals by giving them better access to information. Moreover, in theory, it might appear that a given technology should only expand the choices available to users, since they can abandon it if it creates more problems than it solves. In practice, however, the collective individual decisions made by diplomats using the telegraph sometimes restricted their options and left them with more work, and perhaps worse off, than before the adoption of the technology.[4] Certainly they could now receive much faster answers to their questions. But having faster access to information invited them to act more quickly in moments of stress, and their telegraphic correspondents expected them to provide rapid responses in turn. As a result—and because of the risks of telegraphic espionage and garbling—telegraphy accelerated the pace of labor and increased workloads, especially during diplomatic crises. Thus the benefits of telegraphic diplomacy were difficult to disentangle from the limitations it placed upon its consumers. In the end, it was both a tool and a constraint.

The third question is whether a technology tends to promote authoritarian or democratic power structures. In other words, does it concentrate or distribute power among those who use it and are affected by it? Democratic technologies foster horizontal social networks. Authoritarian media are more easily monopolized by the powerful; they promote vertical linkages through which rulers can collect information from their subordinates and transmit directives to them. The political scientist Ithiel de Sola Pool puts it this way: "Freedom is fostered when the means of communication are dispersed, decentralized, and easily available, as are printing presses or microcomputers. Central control is more likely when the means of communication are concentrated, monopolized, and scarce, as are great networks." Most technologies—although tending in one direction or the other—can fit into either category, depending on how they are used. For example, the moveable-type printing press, which had emancipatory effects in Western Europe, served more oppressive ends in a country like tsarist Russia, where the state closely regulated it.[5]

In general, telegraphy fits more closely into the category of authoritarian media. It encouraged foreign ministries to exercise more control over their agents. Within foreign policy institutions, it lent itself to the creation of hierarchy and specialization. Admittedly, the quest of foreign ministries for greater centralized control generated a backlash: some diplomats struggled against this loss of autonomy. But their resistance was only partially successful in preserving their freedom of action.

Fourth, does a technology accelerate existing trends or push history in a new direction? In general, telegraphy furthered tendencies that were already in existence and would have occurred without it, although often to a lesser extent. The consonance of the telegraph with other historical phenomena of its era was not an accident. The invention and spread of telegraphy owed much to the ingenuity and concern with profit that drove other events often grouped together under the term "Industrial Revolution." At the time of the telegraph's invention, businesses avidly sought faster forms of communication. These insistent demands would have been met somehow, whether by electric telegraphy or by some other technology, such as improved transportation and postal services. The telegraph's acceleration of many existing social trends is therefore unsurprising; it owed its existence to these same trends, and its development was interwoven with theirs.[6]

Telegraphy operated in accordance with a number of other factors—economic development, population growth, an increase in social complexity, and an expansion in the scope of government—to shape foreign ministries. For example, all of these factors contributed to such diplomatic trends as greater workloads, higher expenditures, specialization, bureaucratization, and nascent professionalization. Yet in other respects telegraphic diplomacy did mark a break with the past. The ease with which intelligence services intercepted telegrams produced a new epoch in diplomatic espionage. The adoption of telegraphy enormously accelerated the pace of crisis diplomacy. The telegraph was more prone to garbling than earlier or later media for communicating written language.

But even a problem like garbling was not solely a function of technology. It also resulted from conscious choices made by those who profited from and used the telegraph, many of whom were willing to tolerate higher risks of error for the sake of lower prices and faster transmission. As the problem of garbling demonstrates, we can better assess the influence of a technology if we consider technical considerations (especially the uniqueness of its capabilities and problems) while also taking account of the aspirations of merchandisers and consumers. If we can comprehend the scientific principles, economic pressures, and consumer desires that interact around a given technology, we have an excellent basis from which to estimate its influence. Reasoned inquiry into technological change is imperative. Technology will play a major part in shaping the future of the human species. Indeed, it may determine whether human beings have a future.

Notes

Abbreviations

AA-PA	Auswärtiges Amt, Politisches Archiv (formerly Bonn, now Berlin)
BA	Materials formerly at the Bundesarchiv in Koblenz, Germany, and now in Berlin
BA-MA	Bundesarchiv-Militärarchiv, Freiburg
LC	Library of Congress, Washington, D.C.
MAE, Nantes	Archive of the Ministère des Affaires Étrangères, Nantes
MAE, Paris	Archive of the Ministère des Affaires Étrangères, Paris
NARA	National Archives and Records Administration, College Park, Md., and Washington, D.C.
NMAH	National Museum of American History Archive Center, Smithsonian Institution, Washington, D.C.
PRO	Public Record Office, Kew Gardens, U.K.

Introduction

1. B. R. Mitchell, *International Historical Statistics: Europe, 1750–1993,* 4th ed. (London: Macmillan, 1998), 753–756; B. R. Mitchell, *International Historical Statistics: The Americas, 1750–1993,* 4th ed. (London: Macmillan, 1998), 611.

2. Thomas C. Jepsen, *My Sisters Telegraphic: Women in the Telegraph Office, 1846–1950* (Athens: Ohio University Press, 2000), 190–192; Edwin Gabler, *The American Telegrapher: A Social History, 1860–1900* (New Brunswick: Rutgers University Press, 1988); Jacquelyn Dowd Hall, "O. Delight Smith's Progressive Era," in *Visible Women: New Essays on American Activism,* ed. Nancy A. Hewitt and Suzanne Lebsock (Urbana: University of Illinois Press, 1993); "Telegraph History," *www.chss.montclair.edu/~pererat/perkhx.html.*

3. Historians have recommended studying technological change in order to better understand the conduct of international relations during the nineteenth and twenti-

eth centuries. See Walter LaFeber, "Technology and U.S. Foreign Relations," *Diplomatic History* 24, no. 1 (Winter 2000); Ernest R. May, *American Imperialism: A Speculative Essay* (Chicago: Imprint Publications, 1991), xxxii; James A. Field Jr., "American Imperialism: The Worst Chapter in Almost Any Book," *American Historical Review* 83, no. 3 (June 1978), 660.

4. Elsewhere in the world, it ended later. The United States, for example, did not permanently join this grid until the successful completion of a transatlantic cable in 1866.

5. Mitchell, *International Historical Statistics: Europe,* 673–674, 3–8; Mitchell, *International Historical Statistics: The Americas,* 539.

6. Alexander J. Field, "French Optical Telegraphy, 1793–1855: Hardware, Software, Administration," *Technology and Culture* 35 (1994), 315–347. Frederick Merk, *Manifest Destiny and Mission in American History* (New York: Vintage, 1966), 55–57. Daniel Headrick, *The Invisible Weapon: Telecommunications and International Politics, 1851–1945* (New York: Oxford University Press, 1991), 12.

7. See Jeffrey Kieve, *The Electric Telegraph: A Social and Economic History* (Newton Abbot, U.K.: David and Charles, 1973), 268; Headrick, *Invisible Weapon,* 6.

8. John Bigelow to W. H. Seward, #384, 8 Nov. 1866, M34, #64, NARA. In the Anglo-American world, it was commonplace to speculate on how the latest improvement in telegraphic communication might contribute to making the world more pacific. See statements by Queen Victoria and President Andrew Johnson, 27 and 31 July 1866, folder 1, box 1, ser. 5, collection #73, NMAH; Secretary of State William Seward to the Diplomatic Officers of the U.S. in South America, 18 Aug. 1864, vol. 1, E726 (Circulars of the Department of State), RG 59, NARA; Prime Minister William Gladstone, in *Proceedings at the Banquet Given by Mr. Cyrus W. Field, at the Palace Hotel, Buckingham Gate, London, on Thursday, the 28th November, 1872* (London: R. Clay, Sons, and Taylor, 1872), 2, box 142, collection #205, 1993 addendum, NMAH; President William McKinley to William II, telegram, 30 Aug. 1900, M336, reel #126, NARA.

9. See Charles S. Maier, "Consigning the Twentieth Century to History: Alternative Narratives for the Modern Era," *American Historical Review* 105, no. 3 (June 2000), 816, 820.

10. The Franco-Prussian dispute over the throne of Spain erupted on 3 July 1870; by 15 July war had become inevitable. The publication of the Ems telegram, Otto von Bismarck's carefully edited version of a meeting between the Prussian king and the French ambassador, injured French pride and was the immediate cause of hostilities.

11. Dennis Showalter, *Railroads and Rifles* (Hamdon, Conn.: Archon, 1975). William L. Langer, *European Alliances and Alignments, 1871–1890* (New York: Knopf, 1950), 6.

12. Radio led a boom in technology stocks during the late 1920s that resembled the boom in internet stocks during the 1990s.

13. S. Frederick Starr, "New Communications Technologies and Civil Society,"

in *Science and the Soviet Social Order,* ed. Loren R. Graham (Cambridge, Mass.: Harvard University Press, 1990), 28–29. Ernest R. May, "The News Media and Diplomacy," in *The Diplomats, 1939–1979,* ed. Gordon A. Craig and Francis L. Loewenheim (Princeton: Princeton University Press, 1994), 686. Gordon A. Craig and Alexander L. George, *Force and Statecraft: Diplomatic Problems of Our Time* (New York: Oxford University Press, 1990), 87–99.

14. Mitchell, *International Historical Statistics: The Americas,* 611; Mitchell, *International Historical Statistics: Europe,* 757–760. The economic instability of this era also contributed to the problems of the telegraph industry. Kieve, *Electric Telegraph,* 253.

15. Headrick, *Invisible Weapon,* 198. Kieve, *Electric Telegraph,* 249, 256; Gabler, *American Telegrapher,* 52–53, 133; W. Fred Cottrell, *The Railroader* (Stanford: Stanford University Press, 1940), 21, 55; Jepsen, *Sisters Telegraphic,* 34–35, 190–193.

16. "Phone Diplomacy Arouses Belgians," *New York Times,* 29 June 1931, 10.

17. See David Kahn, *The Codebreakers: The Story of Secret Writing* (New York: Scribner, 1996). The Soviets, although more successful than Britain and the United States at employing human spies, lagged in the collection of signals intelligence. Christopher Andrew and Oleg Gordievsky, *KGB: The Inside Story* (New York: Harper Perennial, 1990), 304–308.

18. Glenn Porter, *The Rise of Big Business, 1860–1920* (Arlington Heights, Ill.: Harlan Davidson, 1992), 41–45. Charles S. Maier, *Dissolution: The Crisis of Communism and the End of East Germany* (Princeton: Princeton University Press, 1997), 93.

19. However, foreign radio broadcasts received in the USSR, such as the those of the BBC and Radio Liberty, played a role in undermining the Soviet regime. Starr, "New Communications Technologies," 23–35.

20. Ibid., 32, 40. Directory assistance was sometimes available. Ithiel de Sola Pool, *Forecasting the Telephone: A Retrospective Technology Assessment* (Norwood, N.J.: Ablex, 1983), 7.

21. Seymour Goodman, "Information Technologies and the Citizen: Toward a 'Soviet-Style Information Society'?" in *Science and the Soviet Social Order,* ed. Graham, 371–372n10, 373n13, 53–54.

22. In the long run, the abandonment of its attempts at economic modernization would probably also have been inconsistent with Soviet attempts to remain a superpower.

23. Leslie David Simon, "The Net: Power and Policy in the 21st Century," in *The Global Century: Globalization and National Security,* ed. Richard L. Kugler and Ellen L. Frost (Washington: National Defense University Press, 2001), 615. Statistics on internet usage from *www.nua.ie/surveys,* esp. "How Many Online" estimate of Aug. 2001, "IDC Research: Half of Europeans Now Online" of 30 Jan. 2002, and "MSNBC: 54 Percent of US Now Online" of 5 Feb. 2002.

24. Andy McSmith, "Sun Sets on 150 Years of the Foreign Office Cable," *Observer,* 16 Aug. 1998, 6. "Department of State Information Technology Strategic Plan, 2001–2005," 3, *www.state.gov/www/dept/irm/strat_plan/ITSP-FContents.html,*

accessed 13 June 2002. The precise end of the telegraph era is difficult to predict. The Foreign Affairs Ministry of New Zealand announced in 1998 that it would continue to use telegrams for important messages because some of its overseas posts were not coupled to its computer system. Simon Kilroy, "E-mail to Replace British Diplomatic Cable," *Dominion* (Wellington), 1 Sept. 1998, 8.

1. The Anglo-American Crisis of 1812

1. See Bradford Perkins, *Prologue to War: England and the United States, 1805–1812* (Berkeley: University of California Press, 1961), 431–432, 436; Irving Brant, *James Madison: Commander in Chief, 1812–1836* (New York: Bobbs-Merrill, 1961), 6: 16–17; Roger H. Brown, *The Republic in Peril: 1812* (New York: Columbia University Press, 1964), 37.

2. Fourth Annual Message, 4 Nov. 1812, in *The Writings of James Madison,* ed. Gaillard Hunt, vol. 8 (New York: Putnam, 1908), 230. Brown, *Republic in Peril.*

3. Leland R. Johnson, "The Suspense Was Hell: The Senate Vote for War in 1812," *Indiana Magazine of History* 65 (1969), 263; Perkins, *Prologue to War,* 411–415, 399.

4. Augustus John Foster, *Jeffersonian America* (San Marino: The Huntington Library, 1954), 100; Perkins, *Prologue to War,* 414–415. "Part of Journal in the United States of America, 1811–1812," entries for 15 June and 30 Jan. 1812, Augustus Foster Papers, LC.

5. See, e.g., Perkins, *Prologue to War,* 301; Denis Gray, *Spencer Perceval: The Evangelical Prime Minister, 1762–1812* (Manchester: Manchester University Press, 1963), 169–170, 177.

6. *Hansard,* XXIII, 542. On 17 June Castlereagh sent a dispatch instructing Foster to inform the U.S. government of the revocation of the Orders. This message, which might have prevented war, did not reach Washington until 5 August. Castlereagh to Foster, #20, 17 June 1812, and enclosure, in Bernard Mayo, ed., *Instructions to the British Ministers to the United States, 1791–1812* (Washington: GPO, 1941), 381–383. The Washington *National Intelligencer* first reported the revocation of the Orders on 6 August 1812. The first official news of the U.S. war declaration reached London on 30 July. See the *Times* (London), 31 July 1812, 2.

7. Foster's correspondence to Castlereagh from 1 April until 17 June 1812 required an average of 37.7 days to get from Washington to the Foreign Office in London; see FO 5/85–86, PRO. From the other direction, according to the Washington *National Intelligencer's* records of ships sailing from England to the United States, 27.5 days was the average journey time for the period from April 1812 until the American declaration of war (the completeness of this information is questionable).

8. The following historians, among others, have explicitly suggested that transatlantic telegraphy would have prevented war: Bernard Mayo, "Introduction," in Mayo, ed., *Instructions to British Ministers,* xv; Johnson, "Suspense Was Hell," 264; Ronald L. Hatzenbuehler and Robert L. Ivie, *Congress Declares War: Rhetoric, Leadership, and Partisanship in the Early Republic* (Kent, Oh.: Kent State University Press,

1983), 115. Other historians have argued that timely American knowledge of the suspension of the Orders would have prevented war: e.g., Perkins, *Prologue to War,* 300, 399, 421; Reginald Horsman, *The Causes of the War of 1812* (Philadelphia: University of Pennsylvania Press, 1962), 254, 260; Brown, *Republic in Peril,* 39–40.

9. After a British attack on an American warship, the *Chesapeake,* in 1807, slow communication resulted in delays while President Jefferson collected information and presented American demands to Britain. Britain refused to meet most of the demands, but in the interim the war fever of the U.S. public dissipated. If the refusal had occurred earlier, when the American public was infuriated with Britain, war would have been much more likely (see Chapter 4). On the second occasion, faster communications would have accelerated the outbreak of war by only a matter of weeks. Madison was determined to commence hostilities unless Britain made major concessions regarding the Orders in Council. In December 1811 the U.S. government, as a final step before war, sent a naval ship, the *Hornet,* to bring back the news from Europe. The trip lasted longer than expected; the *Hornet* did not return until 19 May 1812, and its dispatches reached Washington on 22 May. They contained evidence (inaccurate, as it turned out) that Britain would not repeal the Orders. Madison sent his war message to Congress about a week later. See Perkins, *Prologue to War,* 399–400; Horsman, *Causes of the War of 1812,* 260.

10. Perkins, *Prologue to War,* 432–434; Brown, *Republic in Peril,* 67–68. Journal entry, 18 Apr. 1830, MS Sparks 141g, Jared Sparks MSS, Houghton Library, Harvard University.

11. Anthony Baker, "Minute of a Conversation with Mr. Monroe on Friday July 3, 1812," FO 5/86, PRO; Foster, *Jeffersonian America,* 97; Brown, *Republic in Peril,* 37–38; Jonathan Russell to Joel Barlow, draft, 28 June 1812, Jonathan Russell Papers, Brown University; letter to Abigail Adams, 10 Aug. 1812, in *Writings of John Quincy Adams,* ed. Worthington Chauncey Ford, vol. 4 (New York: Macmillan, 1914), 388.

12. The notion that transatlantic communications were "slow" during this era is valid only in comparison with later years. One should remember that ships sailed at least as fast in 1812 as during any previous time.

13. *Times* (London), 21 July 1812, 3.

14. Henri Hauser, *Economie et diplomatie: Les conditions nouvelles de la politique étrangère* (Paris: Librairie du Recueil Sirey, 1937), 1–2; Neil Hart, *The Foreign Secretary* (Lavenham, Suffolk: Terence Dalton, 1987), 71; minute by Foster, 23 June 1812, 283, FO 5/86, PRO.

15. Twila M. Linville, "The Public Life of Jonathan Russell" (Ph.D. diss., Kent State University, 1971), 139–140; Perkins, *Prologue to War,* 310–312, 390.

16. Otis Ammidon to Russell, 14 July 1811, and Russell to Ammidon, 16 Sept. 1811, Russell Papers. Russell to Monroe, 7 Feb. (private), 13 Feb. (private), 19 Feb., 21 Feb., 4 Mar., 20 Mar., 28 Mar., 9 Apr., 16 Apr., 19 Apr., and 1 May 1812, dispatches, Great Britain, M30, reel 14, NARA; Washington *National Intelligencer,* 6 June 1812. Madison wrote to Thomas Jefferson on 3 April: "A late arrival from G.B. brings dates subsequent to the maturity of the Prince Regent's authority. It appears

that Percival, &c, are to retain their places, and that they prefer war with us, to a repeal of their Orders in Council. We have nothing left therefore, but to make ready for it." *Writings of James Madison,* 8: 185.

17. Russell to Monroe, 8 June 1812, and Russell to Monroe, 9 May 1812, M30, reel 14, NARA. The State Department did not record the arrival date of Russell's 9 May dispatch, but as of 18 June, the date of the U.S. war declaration, the *National Intelligencer* had received news from England only up to the first of May. Moreover, Russell's (9 May) dispatch probably reached Washington no earlier than the news of Perceval's (11 May) assassination, which Madison learned about on 19 June. See J. C. A. Stagg, *Mr. Madison's War: Politics, Diplomacy, and Warfare in the Early American Republic, 1783–1830* (Princeton: Princeton University Press, 1983), 118; *The Republic of Letters: The Correspondence between Thomas Jefferson and James Madison 1776–1826,* ed. James Morton Smith (New York: Norton, 1995), 3: 1681.

18. Russell to Madison, 2 July 1812, Russell Papers. Russell, hoping to be recalled, asserted that both he and Foster were relatively incompetent as diplomats. Remarking about the transfer of a subject of Anglo-American dispute from London to Washington, he declared, "the discussion . . . will undoubtedly be conducted there [Washington] with more ability on our side and less on that of our opponents than here [London]." Russell to Monroe, 9 May 1812, M30, reel 14, NARA.

19. Letter of 15 Feb. 1811, in *The Two Duchesses,* ed. Vere Foster (London: Blackie, 1898), 348.

20. Lady Elizabeth Foster, Duchess of Devonshire, to Foster, 30 Sept. 1805, and Foster to Lady Elizabeth Foster, 2 June 1805, in *Two Duchesses,* 242, 226. Horsman, *Causes of the War of 1812,* 49. Foster to Lady Elizabeth Foster, 30 July 1805, in *Two Duchesses,* 233; Foster, *Jeffersonian America,* 55–56.

21. Foster wrote to the British foreign secretary on 25 Dec. 1811: "It is the opinion of most of the sensible men here that this government will not be pushed into a war with us, but that their object is to secure the support of their party at the next election." Johnson, "Suspense Was Hell," 252n27. Even after the United States declared war, Federalists treacherously told Foster that Britain must "vigorously" prosecute the war in order to bring Madison's government "to a line of conduct reconcileable with Common Sense, and at once produce a solid and advantageous peace." Foster to Castlereagh, unnumbered, 25 Aug. 1812, FO 5/86, PRO.

22. Lady Elizabeth Foster to Foster, 15 Feb. 1811, in *Two Duchesses,* 348; Leslie A. Marchand, *Byron: A Portrait* (New York: Knopf, 1970), 120; Ethel Colburn Mayne, *The Life and Letters of Anne Isabella Lady Noel Byron* (New York: Scribner, 1929), 25. Foster to Lady Elizabeth Foster, 18 Apr. 1812, and Foster, 26 May 1812, in *Two Duchesses,* 360, 365.

23. Foster to Castlereagh, #30, 24 Apr. 1812, FO 5/85, PRO; Perkins, *Prologue to War,* 389. Castlereagh did not respond to Foster's request for a leave of absence; see FO 115/23 and FO 5/83, PRO. Foster to Castlereagh, unnumbered, 25 Aug. 1812, FO 5/86, PRO.

24. Foster to Castlereagh, 15 June 1812, and Foster to Vice-Admiral Sawyer, 15

June 1812, FO 5/86, PRO. Even after the declaration of war, Foster's dispatches showed astounding misjudgments, apparently resulting from his discussions with "the party friendly to Peace with England." Foster reported that "four fifths of the people of the United States are certainly averse to the measures of the Government" and predicted that "New York will declare itself neutral." Foster to Castlereagh, 21 June 1812, ibid.

25. *Hansard,* XXIII, 537, 541. As further evidence of Castlereagh's misreading of the situation, his dispatch #10 of 10 April 1812 suggested that Foster should view time as an ally since delaying tactics would defuse American wrath against Britain. Likewise, on 8 June, he denied the imminence of war with the United States. See Mayo, ed., *Instructions to British Ministers,* 368; Wendy Hinde, *Castlereagh* (London: Collins, 1981), 180; Perkins, *Prologue to War,* 334.

26. Brown, *Republic in Peril,* 30–31; Irving Brant, *James Madison: The President, 1809–1812* (Indianapolis: Bobbs-Merrill, 1956), 371–374; Stagg, *Mr. Madison's War,* 78–80.

27. Brant, *Madison: The President,* 368–369. Foster and other British observers were willing to believe the administration's claims of pacific intentions because, as the *Times* put it, "We have been threatened, and forgiven, and threatened again, in every successive meeting of Congress for several years." *Times* (London), 5 June 1812, 3. As recently as February 1812, U.S. war measures had led Foster to report the imminence of war. This false war scare made Foster more skeptical of later belligerent noises by the United States. See FO 5/84–85, PRO, esp. Foster to Wellesley, 26 and 29 Feb. 1812, #10 and #11. The British had learned to discount American menaces, adopting the explanation advanced by the Federalists, that the threats were mere political posturing designed for a domestic constituency.

28. Foster to Castlereagh, #35, 15 May 1812, FO 5/86, PRO; Perkins, *Prologue to War,* 386. Madison to Jefferson, 3 Apr. 1812, in *Republic of Letters,* 3: 1691. Foster, *Jeffersonian America,* 91–92; Foster, "Part of Journal in the United States of America," 3 Apr. 1812; Foster to Wellesley, #23, 3 Apr. 1812, FO 5/85, PRO.

29. Foster to Wellesley, #22, 2 Apr. 1812, FO 5/85, PRO; Foster to Castlereagh, #35, 15 May 1812, and Foster to Castlereagh, unnumbered, 6 June 1812, FO 5/86, PRO.

30. My argument here has been influenced by political science scholarship about situations in which statesmen believe that an advantage accrues to whichever side first uses military force. According to this scholarship, states that perceive a "first mover advantage" in initiating hostilities are more likely to conceal their grievances, intentions, capabilities, and perceptions, thus impeding effective diplomacy. They are also more likely to engage in hasty or truncated diplomacy. See Stephen Van Evera, *Causes of War: Power and the Roots of Conflict* (Ithaca: Cornell University Press, 1999), 39–53; James D. Fearon, "Rationalist Explanations for War," *International Organization* 49, no. 3 (Summer 1995), 395–396.

31. Madison delivered the speech on 5 November 1811. See Brown, *Republic in Peril,* 89–90, 32; Perkins, *Prologue to War,* 296–297.

32. *Annals of Congress,* 12th Cong., 1st sess., col. 1059. Ironically, in March 1812 Foster secretly reported that the United States sought for Britain "to strike the first Blow and thereby create the Irritation absolutely necessary to enable this Government to raise an army." Foster to Wellesley, 6 Mar. 1812, "Separate and Secret," FO 5/84, PRO. Brant, *Madison: The President,* 375, 367. Monroe to Russell, 5 May 1812, M77, reel 2, NARA; Russell to Monroe, 8 June 1812, M30, reel 14, NARA.

33. Foster to Castlereagh, #49, 24 June 1812, unnumbered, 25 Aug. 1812, and #47, 20 June 1812, FO 5/86, PRO. Likewise, Ralph Ketcham writes that Madison refused to accept Foster's offer of an armistice, which "would forfeit the vital American military advantage of swift action for perhaps meaningless assurances." Ketcham, *James Madison: A Biography* (New York: Macmillan, 1971), 535. Foster's messages to Castlereagh during the period May 21–June 17 required an average of 35.4 days to arrive whereas those from June 18–July 5 averaged 60.6 days, with the *fastest* taking 57 days. See FO 5/86, PRO.

34. Castlereagh's career as foreign secretary showed a pronounced desire to conciliate the United States. Moreover, he tended to subordinate other matters to the war against France, and it is difficult to see how a war with America would have served this cause. However, British statesmen were loath to concede in the face of menacing U.S. behavior. American threats of war, especially if public and truculent, might have made them less likely to repeal the Orders in Council. See Perkins, *Prologue to War,* 205, 206, 216–217.

35. This analysis assumes that other matters would remain the same. A fuller answer to how telegraphy would have affected the use of deceptive diplomacy requires an analysis of complicated military issues (such as how electric telegraphy altered perceived balances between offense and defense in war planning) that are beyond the diplomatic focus of this book.

36. In this instance, I define "poor communication" as a technology whose performance (especially message delivery time) is closely related to the distances involved. Other definitions will be more appropriate for other eras, and may lay greater emphasis upon cost, message integrity, message security, or bandwidth.

2. Diplomatic Autonomy and Telecommunications

1. On telegraphy as a control technology, see Charles S. Maier, "Consigning the Twentieth Century to History: Alternative Narratives for the Modern Era," *American Historical Review* 105, no. 3 (June 2000), 816, 820; James W. Carey, *Communication as Culture: Essays on Media and Society* (Boston: Unwin Hyman, 1989), 212.

2. Edward John Phelps, *International Relations: Address before the Phi Beta Kappa Society of Harvard University, June 29, 1889* (Burlington, Vt.: Free Press Association, 1889), 11. Compare this statement to that of a Tennessee congressman from the 1790s: "Situated as the United States are, at a great distance from the transatlantic world . . . , what have we to do with the politics of Europe?" Warren Frederick Ilchman, *Professional Diplomacy in the United States, 1779–1939: A Study in Administrative History* (Chicago: University of Chicago Press, 1961), 19.

3. Donald E. Queller, *The Office of Ambassador in the Middle Ages* (Princeton: Princeton University Press, 1967), 226; Keith Hamilton and Richard Langhorne, *The Practice of Diplomacy: Its Evolution, Theory and Administration* (London: Routledge, 1995), 31, 238. Garrett Mattingly, *Renaissance Diplomacy* (London: Jonathan Cape, 1955), 58–60.

4. Queller, *Office of Ambassador,* 228. Castlereagh to Foster, #21, 25 June 1812, FO 115/23, PRO.

5. Alfred D. Chandler Jr., *The Visible Hand: The Managerial Revolution in American Business* (Cambridge, Mass.: Harvard University Press, 1977). Ronald H. Coase, the son of a British telegrapher, argued in 1937: "Changes like the telephone and the telegraph which tend to reduce the cost of organizing spatially will tend to increase the size of the firm." R. H. Coase, "The Nature of the Firm," in *The Nature of the Firm: Origins, Evolution, and Development,* ed. Oliver E. Williamson and Sidney G. Winter (New York: Oxford University Press, 1991), 25.

6. Richard B. DuBoff, "The Telegraph in Nineteenth-Century America: Technology and Monopoly," *Comparative Studies in Society and History* 26 (Oct. 1984), 576. Harold D. Woodman, *King Cotton and His Retainers: Financing and Marketing the Cotton Crop of the South, 1800–1925* (Columbia: University of South Carolina Press, 1990), 267, 273–276. Alexander James Field, "The Magnetic Telegraph, Price and Quantity Data, and the New Management of Capital," *Journal of Economic History* 52 (June 1992), 412n33. JoAnne Yates, "The Telegraph's Effect on Nineteenth Century Markets and Firms," *Business and Economic History,* 2d ser., 15 (1986), 158–160.

7. Edwin Gabler, *The American Telegrapher: A Social History, 1860–1900* (New Brunswick: Rutgers University Press, 1988), 14.

8. A. W. Kinglake, *The Invasion of the Crimea* (New York: AMS Press, 1972), 8: 266–267, 263–264; Lytton Strachey, *Eminent Victorians* (London: Penguin, 1986), 135–136. Rebecca Robbins Raines, *Getting the Message Through: A Branch History of the U.S. Army Signal Corps* (Washington: Center of Military History, 1996) 4; Martin van Creveld, *Technology and War: From 2000 B.C. to the Present* (New York: Free Press, 1989), 41–42, 170. Dennis Showalter, "Weapons and Ideas in the Prussian Army from Frederick the Great to Moltke the Elder," in *Tools of War: Instruments, Ideas, and Institutions of Warfare, 1445–1871,* ed. John A. Lynn (Urbana: University of Illinois Press, 1990), 201–202.

9. P. M. Kennedy, "Imperial Cable Communications and Strategy, 1870–1914," *English Historical Review* 86 (1971), 751; Jean-Claude Allain, "L'indépendance câblière de la France au début du XXe siècle," *Guerres mondiales et conflits contemporains,* no. 166 (Apr. 1992), 117. Cromer to Mallet, 2 Aug. 1907, PRO FO/633/98, 5. George Orwell, *The Lion and the Unicorn: Socialism and the English Genius* (Harmondsworth: Penguin, 1982), 61–62. James Carey has even argued that telegraphy, by centralizing imperial control, allowed a shift from colonialism to imperialism. Carey, *Communication as Culture,* 212.

10. My thinking on this subject owes much to Ithiel de Sola Pool. See his *Technologies of Freedom* (Cambridge, Mass.: Harvard University Press, 1983), 5.

11. John Kenneth Galbraith, *A Life in Our Times: Memoirs* (Boston: Houghton Mifflin, 1981), 436–437. Richard B. Kielbowicz, "News Gathering by Mail in the Age of the Telegraph: Adapting to a New Technology," *Technology and Culture* 28 (1987), 26–41.

12. Graeme Davison, *The Unforgiving Minute: How Australia Learned to Tell the Time* (Melbourne: Oxford University Press, 1993), 59. Menahem Blondheim, *News over the Wires: The Telegraph and the Flow of Public Information in America, 1844–1897* (Cambridge, Mass.: Harvard University Press, 1994), 4. Daniel R. Headrick, *The Invisible Weapon: Telecommunications and International Politics, 1851–1945* (New York: Oxford University Press, 1991), 68.

13. Pool, *Technologies of Freedom,* 93. Harold A. Innis, *The Bias of Communication* (Toronto: University of Toronto Press, 1991), 169–170; Daniel J. Czitrom, *Media and the American Mind: From Morse to McLuhan* (Chapel Hill: University of North Carolina Press, 1982), 18.

14. Hamilton Fish circular, 11 Mar. 1875, 33a–33b, vol. 2, E726 (Circulars of the Department of State), RG 59, NARA. Valerie Cromwell and Zara S. Steiner, "The Foreign Office before 1914: A Study in Resistance," in *Studies in the Growth of Nineteenth-Century Government,* ed. Gillian Sutherland (Totowa, N.J.: Rowman and Littlefield, 1972), 184. Elihu Root, Confidential Circular, 16 Jan. 1909, file no. 17530, M862, reel 997, NARA. Circulaire (signed Lacour), "Justification des dépenses de service," #88, 29 Mar. 1883, comptabilité, cartons, 27, MAE, Paris; Ribot, circulaire, #115, 13 June 1890, comptabilité, cartons, 34, MAE, Paris.

15. Robert V. Kubicek, *The Administration of Imperialism: Joseph Chamberlain at the Colonial Office* (Durham: Duke University Press, 1969), 27–28. Joseph C. Grew, *Turbulent Era: A Diplomatic Record of Forty Years, 1904–1945,* ed. Walter Johnson (Boston: Houghton Mifflin, 1952), 317.

16. Monsieur de Callières, *On the Manner of Negotiating with Princes* (1716; Boston: Houghton Mifflin, 1919), 91, 92.

17. Numerous historians of the British empire have emphasized the power that the "man on the spot" derived from the ability to control the information reaching Whitehall. See, e.g., Kubicek, *Administration of Imperialism,* 115, 32, 97; W. David McIntyre, *The Imperial Frontier in the Tropics, 1865–75: A Study of British Colonial Policy in West Africa, Malaya and the South Pacific in the Age of Gladstone and Disraeli* (London: Macmillan, 1967), 381, 15, 42–43.

18. Raymond A. Jones, *The British Diplomatic Service, 1815–1914* (Gerrards Cross, U.K.: Colin Smythe, 1983), 123. William D. Godsey Jr., *Aristocratic Redoubt: The Austro-Hungarian Foreign Office on the Eve of the First World War* (West Lafayette, Ind.: Purdue University Press, 1999), 198.

19. Queller, *Office of Ambassador,* 212, 209, 226. A famous diplomatic handbook asserted, "Now, no matter how far-seeing a minister may be, it is impossible for him to foresee everything or give such ample and at the same time precise instructions to his negotiators as to guide them in all circumstances which may arise." De Callières, *Negotiating with Princes,* 96.

20. "Protocole réservé, signé à Londres, le 15 juillet, 1840 . . . ," in *The Consolidated Treaty Series,* ed. Clive Parry (Dobbs Ferry, N.Y.: Oceana, 1969), 90: 291; Henry Wheaton, *Elements of International Law* (Philadelphia: Lea and Blanchard, 1846), 306–307.

21. Monteagle Stearns, *Talking to Strangers: Improving American Diplomacy at Home and Abroad* (Princeton: Princeton University Press, 1996), 116. Henry E. Mattox, *The Twilight of Amateur Diplomacy: The American Foreign Service and Its Senior Officers in the 1890s* (Kent, Oh.: Kent State University Press, 1989), 74. For Jefferson's comment, see Eric Clark, *Corps Diplomatique* (London: Allen Lane, 1973), 45.

22. Jefferson quoted in Ralph E. Weber, *United States Diplomatic Codes and Ciphers, 1775–1938* (Chicago: Precedent, 1979), 154. Earl of Clarendon to Lord Stratford de Redcliffe, #133, 8 Oct. 1853, *House of Commons Parliamentary Papers,* 1854, vol. 71 (Eastern Papers, pt. II), 142–143. Sir Herbert Maxwell, *The Life and Letters of George William Frederick Fourth Earl of Clarendon* (London: Edward Arnold, 1913), 2: 17–18.

23. Ian K. Steele, *The English Atlantic, 1675–1740: An Exploration of Communication and Community* (New York: Oxford University Press, 1986), 193, 197–198, 208.

24. Seward to Adams, #843, 13 Feb. 1864, in *Papers Relating to Foreign Affairs; Accompanying the Annual Message of the President to the Second Session Thirty-Eighth Congress* (Washington: GPO, 1865), 1864–1865, pt. 1, 171–172.

25. Martin Duberman, *Charles Francis Adams, 1807–1886* (Stanford: Stanford University Press, 1968), 268. See, e.g., Seward to Adams, #807, 11 Jan. 1864, and Adams to Seward, #584, 28 Jan. 1864, ibid., pt. 1, 77, 114–115.

26. Lord John Russell to Lord Lyons, 28 Dec. 1861, PRO 30/22/96, PRO.

27. See Reginald Horsman, *The Causes of the War of 1812* (Philadelphia: University of Pennsylvania Press, 1962); Bradford Perkins, *Prologue to War: England and the United States, 1805–1812* (Berkeley: University of California Press, 1961).

28. Perkins, *Prologue to War,* 207; Canning to Erskine, #2, 23 Jan. 1809, in Bernard Mayo, ed., *Instructions to the British Ministers to the United States, 1791–1812* (Washington: GPO, 1941), 264–266.

29. Perkins, *Prologue to War,* 218. Canning to Erskine, #12, 30 May 1809, in Mayo, ed., *Instructions to British Ministers,* 276. Horsman, *Causes of the War of 1812,* 153.

30. David M. Goldfrank, *The Origins of the Crimean War* (London: Longman, 1994), 29.

31. Buchanan to Louis McLane, 12 July 1845, draft, *James Buchanan Papers at the Historical Society of Pennsylvania* (Philadelphia, 1974), reel 47. Donald A. Rakestraw, *For Honor or Destiny: The Anglo-American Crisis over the Oregon Territory* (New York: Peter Lang, 1995), 153.

32. Thomas R. Hietala, *Manifest Design: Anxious Aggrandizement in Late Jacksonian America* (Ithaca: Cornell University Press, 1985), 232–233. Frederick Merk, *The Oregon Question: Essays in Anglo-American Diplomacy and Politics* (Cambridge, Mass.: Harvard University Press, 1967), 394, 391. Wilbur Devereux Jones, *The Amer-*

ican Problem in British Diplomacy, 1841–1861 (London: Macmillan, 1974), 49–51; Rakestraw, *Honor or Destiny,* 147–153.

33. John A. Munroe, *Louis McLane: Federalist and Jacksonian* (New Brunswick: Rutgers University Press, 1973), 533–534, 546. Hietala, *Manifest Design,* 234. One-way travel between London and Washington took about three weeks in 1846. Howard Jones and Donald A. Rakestraw, *Prologue to Manifest Destiny: Anglo-American Relations in the 1840s* (Wilmington, Del.: SR Books, 1997), 207.

34. See Zachary Karabell, drafter, and Philip Zelikow, ed., "Prelude to War: US Policy toward Iraq 1988–1990" (Cambridge, Mass.: Kennedy School of Government Case Study, 1994), 26–28; Norman Kempster, "Insider; U.S. Ambassador to Iraq Muzzled by Washington," *Los Angeles Times,* 5 Feb. 1991, 5.

35. Harry Braverman, *Labor and Monopoly Capital: The Degradation of Work in the Twentieth Century* (New York: Monthly Review Press, 1974), 195. Braverman (136) quotes the *International Molders Journal* on "deskilling": "the gathering up of . . . scattered craft knowledge, systematizing it and concentrating it in the hands of the employer and then doling it out again only in the form of minute instructions, giving to each worker only the knowledge needed for the performance of a particular relatively minute task. This process, it is evident, separates skill and knowledge even in their narrow relationship. When it is completed, the worker is no longer a craftsman in any sense, but is an animated tool of the management." But while technological change tends to disempower certain workers, it can empower others. For example, automatic teller machines greatly reduce the need for bank tellers but create a demand for better-paid programmers and technicians. Edward Tenner, *Why Things Bite Back: Technology and the Revenge of Unintended Consequences* (New York: Knopf, 1996), 192.

36. Robert S. Lynd and Helen Merrell Lynd, *Middletown: A Study in American Culture* (New York: Harcourt, Brace and World, 1956), 39–41.

37. David F. Long, *Gold Braid and Foreign Relations: Diplomatic Activities of U.S. Naval Officers, 1798–1883* (Annapolis: Naval Institute, 1988), 13; Hugh Thomas, *The Suez Affair* (London: Weidenfeld and Nicolson, 1967), 158.

38. Deposition of W. H. Seward, 7 July 1870, RG 123 (Records of the U.S. Court of Claims), General Jurisdiction, case file 6151, box 306, NARA. See Ralph E. Weber, *Masked Dispatches: Cryptograms and Cryptology in American History, 1775–1900* (Fort Meade, Md.: Center for Cryptologic History, 1993), ch. 17.

39. Viscount Stratford de Redcliffe, 13 May 1861, in *British Parliamentary Papers, Report from the Select Committee on Diplomatic Service, Reports from Committees: 1861,* vol. 6 (ordered by the House of Commons to be printed, 23 July 1861), 168.

40. Unsigned (prob. Robert Lansing) to Woodrow Wilson, 31 Jan. 1916, 763.72/2364, M367, roll #25, NARA. Elihu B. Washburne to Hamilton Fish, #283, 9 Sept. 1870, M34, reel T71, NARA. David Paull Nickles, "Telegraph Diplomats: The United States' Relations with France in 1848 and 1870," *Technology and Culture* 40, no. 1 (Jan. 1999), 11–13, 12.

41. Gordon A. Craig and Felix Gilbert, "Introduction," in Craig and Gilbert,

eds., *The Diplomats, 1919–1939* (New York: Atheneum, 1963), 1: 4. A. L. P. Tucker, *Sir Robert G. Sandeman, K.C.S.I.: Peaceful Conqueror of Baluchistan* (Lahore: Tariq, 1979), 6. "The Need of a Trained Diplomatic Service," *New York Times,* 29 Apr. 1900, 25.

42. René Rémond, *Les Etats-Unis devant l'opinion française, 1815–1852* (Paris: Armand Colin, 1962), 26; Harold Nicolson, *Diplomacy* (London: Thornton Butterworth, 1939), 75. Even independently minded diplomats generally appreciate information on the views of their government. Stratford Canning despised the lethargy of one foreign secretary, who was too lazy to keep him adequately informed. See Algernon Cecil, "The Foreign Office," in *The Cambridge History of British Foreign Policy, 1783–1919,* ed. A. W. Ward and G. P. Gooch, vol. 3 (Cambridge: Cambridge University Press, 1923), 553. Likewise, Elihu B. Washburne, the U.S. minister to France under President Grant and a man with no small measure of self-confidence and initiative, bemoaned the loss of access to State Department instructions during the Prussian siege of Paris. See Nickles, "Telegraph Diplomats," 10n33.

43. Bertie quoted in Sir Horace Rumbold, *Recollections of a Diplomatist* (London: Edward Arnold, 1902), 1: 111–112. Jones, *British Diplomatic Service,* 138. B. D'Agreval, *Les diplomates français sous Napoléon III* (Paris: E. Dentu, 1872), 4; A Diplomate [Arthur Ponsonby], "Diplomacy as a Profession," *National Review* (London) 35 (Mar. 1900), 101; Cecil, "The Foreign Office," 598.

44. Olney to Solomon B. Griffin, 26 Mar. 1913, Richard Olney Papers, reel 58, LC; Gerald G. Eggert, *Richard Olney: Evolution of a Statesman* (University Park: Pennsylvania State University Press, 1974), 318.

45. J. D. Gregory, *On the Edge of Diplomacy: Rambles and Reflections, 1902–1928* (London: Hutchinson, 1931?), 15. Ole R. Holsti, *Crisis, Escalation, War* (Montreal: McGill-Queen's University Press, 1972), 95–101.

46. Finley Peter Dunne, "Mr. Dooley on Diplomatic Indiscretions," *New York Times,* 22 Mar. 1914, sec. 5, 5.

47. Felix Gilbert, *To the Farewell Address: Ideas of Early American Foreign Policy* (Princeton: Princeton University Press, 1961), 72–74. Dumas Malone, *Jefferson and the Rights of Man* (Boston: Little, Brown, 1951), 401; Weber, *U.S. Diplomatic Codes and Ciphers,* 121.

48. Samuel Sullivan Cox, "The Folly and Cost of Diplomacy," speech of 16 May 1874 in the House of Representatives (Washington, 1874), 4; Ilchman, *Professional Diplomacy in the United States,* 20–21. Ezra Stiles Gannett, *The Atlantic Telegraph: A Discourse Delivered in the First Church, August 8, 1858* (Boston: Crosby, Nichols, 1858), 12. American politicians continue to argue that improved communications make diplomatic representation unnecessary. See *Washington Post,* 3 June 1996, A1, A12; Johanna Neuman, *Lights, Camera, War: Is Media Technology Driving International Politics?* (New York: St. Martin's, 1996), 4.

49. Cox, "Folly and Cost," 4. Question posed to Earl Cowley, 30 May 1861, in *British Parliamentary Papers, Report from the Select Committee on Diplomatic Service, Reports from Committees: 1861,* 6: 232. One encounters the view that telegraphy ren-

dered diplomats unnecessary much less often in the 1870 and 1871 parliamentary investigations into the diplomatic service than in the 1861 investigation.

50. Lord Ampthill to Lord Granville, 12 Mar. 1881, in Paul Knaplund, ed., *Letters from the Berlin Embassy, 1871–1874, 1880–1885* (Washington, 1944), 208. Christoph von Tiedemann, *Sechs Jahre Chef der Reichskanzlei unter dem Fürsten Bismarck* (Leipzig: Hirzel, 1910), 211–213. Moritz Busch, *Our Chancellor: Sketches for a Historical Picture,* trans. William Beatty-Kingston (New York: Scribner, 1884), 1: 245. On Bismarck's subjugation of the Prussian diplomatic corps, see Gordon Craig, "Bismarck and His Ambassadors: The Problem of Discipline," in Craig, ed., *War, Politics, and Diplomacy: Selected Essays* (New York: Praeger, 1966), 187–193; John C. G. Röhl, *The Kaiser and His Court: Wilhelm II and the Government of Germany,* trans. Terence F. Cole (Cambridge: Cambridge University Press, 1987), 158; Karl-Alexander Hampe, *Das Auswärtige Amt in der Ära Bismarck* (Bonn: Bouvier Verlag, 1995), 89–91.

51. Bismarck to Canitz, 1 June 1870, in Georges Bonnin, ed., *Bismarck and the Hohenzollern Candidature for the Spanish Throne* (London: Chatto and Windus, 1957), 166. Lamar Cecil, *The German Diplomatic Service, 1871–1914* (Princeton: Princeton University Press, 1976), 320–323; Röhl, *Kaiser and His Court,* 158–159; Emily Oncken, *Panthersprung nach Agadir: Die deutsche Politik während der Zweiten Marokkokrise 1911* (Düsseldorf: Droste, 1981), 146–147. Bismarck suggested that railroads and telegraphy augmented the ability of foreign ministers to supervise diplomats. George O. Kent, *Arnim and Bismarck* (Oxford: Clarendon, 1968), 65.

52. M. B. Hayne, *The French Foreign Office and the Origins of the First World War, 1898–1914* (Oxford: Clarendon, 1993), 84, 81.

53. Hugh G. J. Aitken, *Scientific Management in Action: Taylorism at Watertown Arsenal, 1908–1915* (Princeton: Princeton University Press, 1985), 169–185, 209–234; David F. Noble, "Social Choice in Machine Design: The Case of Automatically Controlled Machine Tools," in *Case Studies on the Labor Process,* ed. Andrew Zimbalist (New York: Monthly Review Press, 1979), 38–44; David Montgomery, *Workers' Control in America: Studies in the History of Work, Technology, and Labor Struggles* (Cambridge: Cambridge University Press, 1979), 113–116.

54. Susan Douglas, "The Navy Adopts the Radio, 1899–1919," in *Military Enterprise and Technological Change: Perspectives on the American Experience,* ed. Merritt Roe Smith (Cambridge, Mass.: MIT Press, 1985), 148–149. James Q. Wilson, *Bureaucracy: What Government Agencies Do and Why They Do It* (New York: Basic Books, 1989), 168–169. On causes and types of disobedience, see Robert Jervis, *Perception and Misperception in International Politics* (Princeton: Princeton University Press, 1976), 332–338.

55. Nicholas Henderson, *Mandarin: The Diaries of an Ambassador, 1969–1982* (London: Weidenfeld and Nicolson, 1994), 3. See, e.g., Aberdeen to Lord Russell, 16 Sept 1853, folio 226, Lord John Russell Papers, PRO 30/22/11A, PRO. On Stratford's supposed responsibility for the Crimean War, see Kinglake, *Invasion of the Crimea,* 1:

374ff; Norman Rich, *Why the Crimean War? A Cautionary Tale* (Hanover, N.H.: University Press of New England, 1985), 60, 70, 78; R. W. Seton-Watson, *Britain in Europe, 1789–1914: A Survey of Foreign Policy* (1937; New York: Howard Fertig, 1968), 312, 316–318.

56. Aberdeen, 15 Feb. 1853, in Spencer Walpole, *The Life of Lord John Russell* (New York: Haskell, 1969), 2: 178. Seton-Watson, *Britain in Europe,* 318. For one example of purposeful misinterpreting of instructions, see Stanley Lane-Poole, *The Life of the Right Honourable Stratford Canning, Viscount Stratford de Redcliffe* (London: Longmans, Green, 1888), 2: 308. For defenses of Stratford, see Paul W. Schroeder, *Austria, Great Britain, and the Crimean War: The Destruction of the European Concert* (Ithaca: Cornell University Press, 1972), 409–410; Ann Pottinger Saab, *The Origins of the Crimean Alliance* (Charlottesville: University Press of Virginia, 1977), 66–75; Goldfrank, *Origins of the Crimean War,* 276–278.

57. Stratford to Clarendon, #650, 3 Sept. 1855, Stratford Papers, FO 352/54b, PRO. Viscount Stratford de Redcliffe, 13 May 1861, in *British Parliamentary Papers, Report from the Select Committee on Diplomatic Service, Reports from Committees: 1861,* 6: 168. For examples of garbled telegrams Stratford received, see Stratford to Clarendon, #427, 13 Apr. 1856, FO 352/55a, and Stratford to Sir Henry Bulwer, 23 June 1857, FO 352/46, Stratford Papers, PRO.

58. Stratford's remarks on the necessity for independently minded diplomats in the age of the telegraph show great similarity to the comments he made after ignoring his instructions three decades earlier, in 1829: "I should really think myself unworthy . . . of any place whatever in the trust of my sovereign, if I were to shrink from the responsibility of modifying or suspending any part of my instructions, when happening to be in possession of information unknown at the time to H.M.'s Government and calculated in my opinion to affect materially their views of the question involved in those instructions." See Neil Hart, *The Foreign Secretary* (Lavenham: Terence Dalton, 1987), 53.

59. Lord John Russell testimony, 24 June 1861, *British Parliamentary Papers, Report from the Select Committee on Diplomatic Service, Reports from Committees: 1861,* 6: 308. John Tilley and Stephen Gaselee, *The Foreign Office* (London: Putnam's, 1933), 257. As I argued earlier, concision could also have had the opposite effect: concise telegrams often sounded more authoritative than regular dispatches and thus may have encouraged obedience among some diplomats.

60. David Stevenson, *Armaments and the Coming of War: Europe, 1904–1914* (Oxford: Clarendon, 1996), 382. J. F. V. Keiger, *Raymond Poincaré* (Cambridge: Cambridge University Press, 1997), 172. Hayne, *French Foreign Office,* 84. William A. Renzi, "Who Composed 'Sazonov's Thirteen Points'? A Re-Examination of Russia's War Aims of 1914," *American Historical Review* 88, no. 2 (Apr. 1983), 352.

61. Hayne, *French Foreign Office,* 119, 121, 253, 294–295; Stevenson, *Armaments,* 382, 392. Luigi Albertini, *The Origins of the War of 1914* (London: Oxford University Press, 1953), 2: 538; Maurice Paléologue, *An Ambassador's Memoirs* (London: Hutch-

inson, 1923), 1: 32; Peter Herzog, *Glaubwürdigkeit und Quellenwert der Tagebücher des französischen Botschafters Paléologue: Eine kritische Untersuchung* (1940; Darmstadt: Scientia Verlag Aalen, 1982), 50n11.

62. Stevenson, *Armaments,* 388, 381–382; Keiger, *Poincaré,* 174–175; Hayne, *French Foreign Office,* 294–301; John F. V. Keiger, *France and the Origins of the First World War* (London: Macmillan, 1983), 155, 159, 160; Albertini, *Origins of the War,* 2: 295–296, 536–539, 582–588, 602–603, 617–627; William Jannen Jr., *The Lions of July: Prelude to War, 1914* (Novato, Calif.: Presidio, 1996), 83, 86; L. C. F. Turner, "The Russian Mobilization in 1914," *Journal of Contemporary History* 3, no. 1 (1968), 81, 83–85; Pierre Renouvin, "La politique française en juillet 1914 d'après les documents diplomatiques français," *Revue d'Histoire de la Guerre Mondiale* 15, no. 1 (Jan. 1937), 10; Jules Isaac, *Un dèbat historique: Le problème des origines de la guerre* (Paris: Rieder, 1933), 190–214.

63. Keiger, *Poincaré,* 176. Historians have similarly accused Germany's ambassador at Vienna, Count Heinrich von Tschirschky, of sabotaging his government's last-minute effort to restrain Austria-Hungary during the July crisis. For example, because of Tschirschky's poor reporting, his own government learned about Austria-Hungary's general military mobilization long after Russian statesmen did. Albertini, *Origins of the War,* 2: 667–668.

64. Jean Stengers, "1914: The Safety of Ciphers and the Outbreak of the First World War," in *Intelligence and International Relations, 1900–1945,* ed. Christopher Andrew and Jeremy Noakes (Exeter: University of Exeter Press, 1987); Stevenson, *Armaments,* 390; Keiger, *Poincaré,* 174–175.

65. Paléologue to French foreign ministry, telegram 304, 29 July 1914, 11:45 pm, PA-AP/133, vol. 1, MAE, Paris. Stengers suggests that Russian officials withheld information from Paléologue because they thought him indiscreet; "1914," 35–36. But the examples he cites are of brief delays (in the clearest example, about an hour) and were in no way decisive. Moreover, Paléologue never attempted to defend himself by claiming that the Russian government had kept him misinformed; Albertini, *Origins of the War,* 2: 584, 624.

66. See Nicolas de Basily, *Memoirs: Diplomat of Imperial Russia, 1903–1917* (Stanford: Hoover Institution, 1973), 96; Raymond Recouly, *Les heures tragiques d'avant guerre* (Paris: La Renaissance du Livre, 1922?), 160–161; Youri Danilov, *La Russie dans la guerre mondiale, 1914–1917* (Paris: Payot, 1927), 39.

67. Albertini, *Origins of the War,* 2: 625. Stengers, "1914," 36, 43; Basily, *Memoirs,* 96; Recouly, *Les heures tragiques,* 160–161; Danilov, *La Russie dans la guerre mondiale,* 39. Russia's failure to keep France properly informed hindered French military preparations and gave France full legal right not to honor the Franco-Russian military agreement during the July crisis. Albertini, *Origins of the War,* 2: 585.

68. Basily, *Memoirs,* 96. Keiger, *France and the Origins of the First World War,* 154. For other instances of Paléologue's poor reporting prior to 29 July 1914, see Hayne, *French Foreign Office,* 298–299.

69. David Kahn, *The Codebreakers: The Story of Secret Writing* (New York: Scribner, 1996), 621; Renzi, "Sazonov's Thirteen Points," 355.

70. Viviani to Paléologue, "copie," nos. 483–484, 31 July 1914, 9:00 P.M. and 9:20 P.M., PA-AP/133, vol. 1; note by Viviani, 3 Jan. 1923, PA-AP/133, vol. 5, MAE, Paris; Baron Wilhelm von Schoen, *The Memoirs of an Ambassador: A Contribution to the Political History of Modern Times,* trans. Constance Vesey (London: George Allen and Unwin, 1922), 193–194; Albertini, *Origins of the War,* 2: 605; Ulrich Trumpener, "War Premeditated? German Intelligence Operations in July 1914," *Central European History* 9, no. 1 (Mar. 1976), 80–83.

71. Albertini, *Origins of the War,* 2: 622–623.

72. Katherine Anne Lerman, *The Chancellor as Courtier: Bernhard von Bülow and the Governance of Germany, 1900–1909* (Cambridge: Cambridge University Press, 1990), 193; Cecil, *German Diplomatic Service,* 213, 300–301.

73. For an example, see Kinglake, *Invasion of the Crimea,* 8: 266–267, 377n2.

74. Georges Bonnin, "Introduction," in Bonnin, ed., *Bismarck and the Hohenzollern Candidature for the Spanish Throne* (London: Chatto and Windus, 1957), 24.

75. Bethmann Hollweg to Schoen, #193, 3 Aug. 1914, Bethmann Hollweg to Wolff, 7 Aug. 1914, Schoen to Viviani, 3 Aug. 1914, and Schoen to Bethmann Hollweg, 6 Aug. 1914, all at T120, reel 1821 [Bd. 16], NARA. The telegrams can be found in *Outbreak of the World War: German Documents Collected by Karl Kautsky,* ed. Karl Kautsky, Max Montgelas, and Walter Schücking (New York: Oxford University Press, 1924), 530–532.

76. Schoen, *Memoirs,* 200–201.

77. Albertini, *Origins of the War,* 3: 213–215. A. Aulard, "Ma controverse avec Delbrück," *La Revue de Paris,* 1 May 1922, 40–43; Pierre Renouvin, *The Immediate Origins of the War (28th June—4th August 1914),* trans. Theodore Carswell Hume (New Haven: Yale University Press, 1928), 270–271.

78. Renouvin, *Immediate Origins,* 269, 265–266. On Schoen's disputes with French scholars, see, e.g., Harry Elmer Barnes, "The German Declaration of War on France: The Question of Telegram Mutilations; Premier Poincaré Versus Ambassador von Schoen," *American Historical Review* 35, no. 1 (Oct. 1929), 76–78.

79. David Newsome, a U.S. diplomat, made this point to me on 15 Feb. 1997; Stewart Eldon, "Foreign Ministries and the Information Revolution," *World Today* 50, nos. 8–9 (Aug.–Sept. 1994), 177.

80. See Robert J. Cain, "Telegraph Cables in the British Empire, 1850–1900" (Ph.D. diss., Duke University, 1970), 190.

81. Interview with Stewart Eldon, a British diplomat, 26 Apr. 1994.

82. Albertini, *Origins of the War,* 2: 460–461.

83. Alfred P. Sloan Jr., *My Years with General Motors* (Garden City, N.Y.: Doubleday, 1964), 53. And see Thomas K. McCraw, *American Business, 1920–2000: How It Worked* (Wheeling, Ill.: Harlan Davidson, 2000), 7–8, 209–211.

84. For analyses of this problem from the realms of military history, political sci-

ence, and social psychology, see Martin Samuels, *Command or Control? Command, Training and Tactics in the British and German Armies, 1888–1918* (London: Frank Cass, 1995), 283–285; Wilson, *Bureaucracy,* 43–44; Dietrich Dörner, *The Logic of Failure: Why Things Go Wrong and What We Can Do to Make Them Right* (New York: Metropolitan Books, 1996), 161–162.

85. See Thomas K. McCraw and Richard S. Tedlow, "Henry Ford, Alfred Sloan, and the Three Phases of Marketing," in *Creating Modern Capitalism: How Entrepreneurs, Companies, and Countries Triumphed in Three Industrial Revolutions,* ed. McCraw (Cambridge, Mass.: Harvard University Press, 1997); Michael C. Jensen, *Foundations of Organizational Strategy* (Cambridge, Mass.: Harvard University Press, 1998).

86. C. W. de Kiewiet, *The Imperial Factor in South Africa: A Study in Politics and Economics* (Cambridge: Cambridge University Press, 1937), 293.

87. Robert D. Schulzinger, *Henry Kissinger: Doctor of Diplomacy* (New York: Columbia University Press, 1989), 151.

88. Lord Strang et al., *The Foreign Office* (1955; Westport, Conn.: Greenwood, 1984), 117.

3. The Trent Affair

1. C. P. Stacey, *Canada and the British Army, 1846–1871* (Toronto: University of Toronto Press, 1963), 152. Brian Jenkins, *Britain and the War for the Union* (Montreal: McGill-Queen's University Press, 1974), 1: 187.

2. Norman Ferris, *Desperate Diplomacy: William H. Seward's Foreign Policy, 1861* (Knoxville: University of Tennessee Press, 1976), 180–187.

3. Stanley Gallas, "Lord Lyons and the Civil War, 1859–1864: A British Perspective" (Ph.D. diss., University of Illinois at Chicago Circle, 1982), 1, 10–11, 149–150.

4. Wilfrid Ward, "Lord Lyons in Private Life," in Lord Newton, *Lord Lyons: A Record of British Diplomacy* (New York: Longmans, Green, 1913), 2: 417. Eugene H. Berwanger, *The British Foreign Service and the American Civil War* (Lexington: University Press of Kentucky, 1994), 23–24.

5. Walter LaFeber, *The New Empire: An Interpretation of American Expansion, 1860–1898* (1963; Ithaca: Cornell University Press, 1988), 28. Seward seemed to believe that such an annexation was inevitable and would be peaceful; it seems doubtful that he seriously contemplated a war against Britain.

6. *Congressional Globe* (Washington: John C. Rives, 1857), 3rd sess., 34th Cong., 257, 258, 1086; *The Works of William H. Seward,* ed. George P. Baker (Boston: Houghton Mifflin, 1884), 4: 45–46; Charles Vevier, "The Collins Overland Line and American Continentalism," *Pacific Historical Review* 28 (Aug. 1959), 247, 251–252. Seward's support for such projects continued when he was secretary of state. Deposition of W. H. Seward, 7 July 1870, RG 123 (Records of the U.S. Court of Claims), General Jurisdiction, case file 6151, box 306, NARA.

7. Speech of 12 Sept. 1860, in Seward, *Works,* 4: 319.

8. Lyons to Lord John Russell, 7 Jan. 1861, in Newton, *Lord Lyons,* 1: 30. Lyons to Russell, "private," 2 May 1861, PRO 30/22/35, PRO. Lyons to Sir E. Head, 2 Aug. 1861, ibid., 1: 50.

9. Gordon H. Warren, *Fountain of Discontent: The Trent Affair and Freedom of the Seas* (Boston: Northeastern University Press, 1981), 61–63; Glyndon G. Van Deusen, *William Henry Seward* (New York: Oxford University Press, 1967), 281–283; David P. Crook, *The North, the South, and the Powers, 1861–1865* (New York: Wiley, 1974), 57–61.

10. Crook, *North, South, and Powers,* 103–106, 110, 111; Jenkins, *Britain and the War for the Union,* 1: 196–197. The decision to release the *Trent* but retain the diplomats greatly weakened the legal case of the North. Even if Mason and Slidell could be considered contraband, a dubious proposition, only a prize court could decide upon their seizure. My discussion of the Trent affair is informed by Warren's *Fountain of Discontent* and by Norman B. Ferris, *The Trent Affair: A Diplomatic Crisis* (Knoxville: University of Tennessee Press, 1977).

11. David Herbert Donald, *Charles Sumner* (New York: Da Capo, 1996), pt. 2, 31.

12. David Herbert Donald, *Lincoln* (New York: Simon and Schuster, 1995), 322–323; Jenkins, *Britain and the War for the Union,* 1: 202–203. Lyons to Commodore Dunlop, "copy," 9 Dec. 1861, FO 5/776, PRO.

13. Warren, *Fountain of Discontent,* 109. Horace Rumbold, *Recollections of a Diplomatist* (London: Edward Arnold, 1902), volume 2, 83. Lynn M. Case and Warren F. Spencer, *The United States and France: Civil War Diplomacy* (Philadelphia: University of Pennsylvania Press, 1970), 195–205; London Embassy to Ministère des Affaires Étrangères, telegrams, 27 Nov. 1861 and 9 Dec. 1861, Correspondance Politique, Angleterre, 720, MAE, Paris.

14. Warren, *Fountain of Discontent,* 115–117; Ferris, *Trent Affair,* 50–53; Jenkins, *Britain and the War for the Union,* 1: 211–213; Newton, *Lord Lyons,* 1: 62–63.

15. Kenneth Bourne, *Britain and the Balance of Power in North America, 1815–1908* (Berkeley: University of California Press, 1967), 219–247; Jenkins, *Britain and the War for the Union,* 213–218.

16. Ferris, *Trent Affair,* 138; Bourne, *Britain and the Balance of Power,* 244–246. Lyons to Russell, 22 Nov. 1861, and Russell to Lyons, 1 Dec. 1861, in Newton, *Lord Lyons,* 1: 55, 63.

17. Lyons to Russell, 19 Dec. 1861, in Newton, *Lord Lyons,* 1: 66.

18. Lyons to Russell, 23 Dec. 1861, ibid., 67; Warren, *Fountain of Discontent,* 177; Ferris, *Trent Affair,* 132–133.

19. Daniel B. Carroll, *Henri Mercier and the American Civil War* (Princeton: Princeton University Press, 1971), 112–113. Seward, *Works,* 5: 295–309.

20. Steamboats had reduced the travel time between Britain and the United States to a typical range of 12–14 days by 1861. See Warren, *Fountain of Discontent,* 140; Gallas, "Lord Lyons and the Civil War, 1859–1864," 11n13.

21. Menahem Blondheim of the Hebrew University of Jerusalem suggests that

the 1858 cable never worked and was a fraud. Blondheim, "European News from the Cable's Mouth: International, National, and Local Interfaces," manuscript, 10.

22. Samuel Carter III, *Cyrus Field: Man of Two Worlds* (New York: Putnam, 1968), 193; Vary T. Coates, Bernard Finn, et al., *A Retrospective Technology Assessment: Submarine Telegraphy—The Transatlantic Cable of 1866* (San Francisco: San Francisco Press, 1979), 20. Newton, *Lord Lyons,* 1: 77–78. Charles Francis Adams Jr., the son of the American minister to Britain, also contended that a working transatlantic cable would have increased the dangers of the crisis. Adams, *The Trent Affair: An Historical Retrospect* (Boston, 1912), 10–11.

23. Warren, in *Fountain of Discontent,* 219, supports Field. A number of historians have taken the other perspective, e.g., David F. Long, *Gold Braid and Foreign Relations: Diplomatic Activities of U.S. Naval Officers, 1798–1883* (Annapolis: Naval Institute Press, 1988), 325–326; Thomas A. Bailey, *A Diplomatic History of the American People* (New York: Appleton-Century-Crofts, 1964), 329–330; R. B. Mowat, *The Diplomatic Relations of Great Britain and the United States* (New York: Longmans, Green, 1925), 174, 179.

24. *New York Herald,* 29 Dec. 29, 1861, 4. Lyons sent this article to Russell on 31 Dec. 1861; #807, FO 5/777, PRO. The *Herald*'s assertion is dubious since Seward's dispatch of 30 November failed to convince Palmerston that the crisis was over. See Ferris, *Trent Affair,* 149.

25. Sir George C. Lewis to William E. Gladstone, 24 Dec. 1861, *Papers of the Prime Ministers of Great Britain,* ser. 8 (Gladstone Papers), pt. 4, reel 16, Add. Mss. 44236.

26. Bright to Sumner, in Donald, *Charles Sumner,* pt. 2, 37. Crook, *North, South, and Powers,* 116.

27. Warren, *Fountain of Discontent,* 45–46, 144, 152, 167; Ferris, *Trent Affair,* 35–36, 156–158; Donald, *Lincoln,* 322; Donald, *Charles Sumner,* pt. 2, 37.

28. Robert Jervis, "Hypotheses on Misperception," in Robert J. Art and Robert Jervis, eds., *International Politics: Enduring Concepts and Contemporary Issues,* 3rd ed. (New York: HarperCollins, 1992), 483; Robert Jervis, *Perception and Misperception in International Politics* (Princeton: Princeton University Press, 1976), 62–67.

29. Lyons to Hammond, in Newton, *Lord Lyons,* 1: 77. Warren, *Fountain of Discontent,* 34. Adams, *Trent Affair,* 5–6.

30. Coates and Finn, *Retrospective Technology Assessment,* 83, 91.

31. Van Deusen, *William Henry Seward,* 370–371. Berwanger, *British Foreign Service and the American Civil War,* 24–25.

32. Russell wrote that he had considered giving Lyons "contingent instructions. But the result is that I fear I should embarrass you more by such a course than by leaving you to the exercise of your own excellent judgement." Russell to Lyons, 7 Dec. 1861, PRO 30/22/96, PRO. Adams's diary displays the self-possession and good judgment with which he received news of the *Trent* affair; *Adams Papers,* pt. 1, reel 76, Massachusetts Historical Society, 1954.

33. Russell to Lyons, 11 Jan. 1862, in Raymond A. Jones, *The British Diplomatic Service, 1815–1914* (Gerrards Cross, U.K.: Colin Smythe, 1983), 131. Wilfrid Ward, "Lord Lyons in Private Life," in Newton, *Lord Lyons,* 2: 425.

4. Speed and Diplomacy

1. Nathaniel Hawthorne's fictional character Clifford Pyncheon imputed "an immaterial and miraculous power" to the telegraph. Hawthorne, *The House of the Seven Gables and The Snow Image and Other Twice Told Tales* (1851; Boston: Houghton Mifflin, 1883), 314. For similar examples from Bram Stoker's *Dracula,* see Jay Pawlowski, "A Daemon in Her Shape: Dracula and Nineteenth-Century Communication Technologies," *Antenna: Newsletter of the Mercurians, in the Society for the History of Technology* 13, no. 1 (Nov. 2000), 8–9. Visual means of communication, such as smoke signals or optical telegraphy, and aural means, such as bells or drums, also separated communication from transportation. But these media were slower, possessed less carrying capacity, and were more dependent on favorable weather than the electric telegraph.

2. James Carey, *Communication as Culture: Essays on Media and Society* (Boston: Unwin Hyman, 1989), 203, 212.

3. Daniel J. Czitrom, *Media and the American Mind: From Morse to McLuhan* (Chapel Hill: University of North Carolina Press, 1982), 7. Annteresa Lubrano, *The Telegraph: How Technology Innovation Caused Social Change* (New York: Garland, 1997), 125. Lester G. Lindley, *The Impact of the Telegraph on Contract Law* (New York: Garland, 1990), 25. Electric telegraphy was one of many nineteenth-century technologies and institutions (such as canals, the post office, railroads) that enthralled a public excited by speed and efficiency. See Leo Marx, *The Machine in the Garden: Technology and the Pastoral Ideal in America* (New York: Oxford University Press, 1970), 194, 230; Richard R. John, "American Historians and the Concept of the Communications Revolution," in *Information Acumen: The Understanding and Use of Knowledge in Modern Business,* ed. Lisa Bud-Frierman (London: Routledge, 1994).

4. Ernest R. May, *American Imperialism: A Speculative Essay* (Chicago: Imprint Publications, 1991), 193. Tom Standage, *The Victorian Internet: The Remarkable Story of the Telegraph and the Nineteenth Century's On-line Pioneers* (New York: Walker, 1988), 102. Adams quoted in Stewart Brand, *The Clock of the Long Now: Time and Responsibility* (New York: Basic Books, 1999), 16. Paul Pradier-Fodéré, *Cours de droit diplomatique* (La Rochelle: Noel Texier, 1899), 1: vii.

5. Daniel Headrick, *The Invisible Weapon: Telecommunications and International Politics, 1851–1945* (New York: Oxford University Press, 1991), 199–201; Vary T. Coates, Bernard Finn, et al., *A Retrospective Technology Assessment: Submarine Telegraphy: The Transatlantic Cable of 1866* (San Francisco: San Francisco Press, 1979), 27–30.

6. Washburne to Hamilton Fish, #283, 9 Sept. 1870, M34, reel T71, NARA.

7. See Robert Howard Lord, *The Origins of the War of 1870: New Documents from*

the German Archives (Cambridge, Mass.: Harvard University Press, 1924), 81–83; Luigi Albertini, *The Origins of the War of 1914*, vol. 2 (Oxford: Oxford University Press, 1953), 525n6, 610–611, 617–618.

8. Testimony of the Earl of Clarendon, 16 June 1870, *British Parliamentary Papers, Report from the Select Committee on Diplomatic and Consular Services* (ordered by the House of Commons to be printed, 25 July 1870), 7: 293. Bryan to Gerard, 23 May 1915, 763.72/2440a, M367, reel 25, NARA. During the Second World War, the process of coding messages often produced a two-day delay and sometimes as long as five to seven days. Coates and Finn, *Retrospective Technology Assessment,* 87.

9. See, e.g., Albertini, *Origins of the War,* 2: 504. During the First World War the German ambassador to Constantinople asked his government to consider sending some messages in clear in order to speed up their reception. Kühlmann to Bethmann Hollweg, 13 Jan. 1917, R2444, AA-PA.

10. Graeme Davison, *The Unforgiving Minute: How Australia Learned to Tell the Time* (Melbourne: Oxford University Press, 1993), 59. Manager of the Postal Telegraph-Cable Company to the German embassy in Washington, 5 Mar. 1904, Washington Botschaft 334, AA-PA. E. J. Hale, 27 Nov. 1913, 119.2/62, 1910–29, RG 59, NARA. Andy McSmith, "Sun Sets on 150 Years of the Foreign Office Cable," *Observer,* 16 Aug. 1998, 6.

11. David Kahn, *The Codebreakers: The Story of Secret Writing* (New York: Scribner, 1997), 63–64.

12. Howard W. French, "Pearl Harbor Truly a Sneak Attack, Papers Show," *New York Times,* 9 Dec. 1999, A5. Tai Kawabata, "What Was Fair in War Not a Question of Timing: Documents Suggest Memo Meant to Deceive U.S., Keep Pearl Harbor Attack Secret," *Japan Times,* 7 Dec. 1999, 3; Kahn, *Codebreakers,* 43, 59, 60, 62–63.

13. Frelinghuysen to Trescot, #6, 9 Jan. 1882, and Trescot to Frelinghuysen, #8, 3 Feb. 1882, in *Foreign Relations of the United States: 1882* (Washington: GPO, 1883), 57–58, 67–68; Russell H. Bastert, "A New Approach to the Origins of Blaine's Pan American Policy," *Hispanic American Historical Review* 39 (Aug. 1959), 408–409.

14. Count Alexandre Walewski to French Foreign Ministry, 17 Nov. 1851, Affaires diverses politiques, Angleterre, 12, MAE, Paris. Hammonds to Lyons, "private," 9 Feb 1870, FO 391/13, PRO.

15. *The Paris Embassy during the Second Empire: Selections from the Papers of Henry Richard Charles Wellesley, 1st Earl Cowley,* ed. F. A. Wellesley (London: Thornton Butterworth, 1928), 24.

16. Richard M. Sorrentino and Neil Vidmar, "Impact of Events: Short- vs. Long-term Effects of a Crisis," *Public Opinion Quarterly* 38 (Summer 1974), 272, 279. John E. Mueller, *War, Presidents and Public Opinion* (Lanham, Md.: University Press of America, 1985), 267. Jong R. Lee, "Rallying around the Flag: Foreign Policy Events and Presidential Popularity," *Presidential Studies Quarterly* 7 (Fall 1977), 253. Larry Hugick and Alec M. Gallup, "'Rally Events' and Presidential Approval," *Gallup Poll Monthly,* June 1991, 24–25. May, *American Imperialism.* Richard A. Brody, *Assessing*

the President: The Media, Elite Opinion, and Public Support (Stanford: Stanford University Press, 1991), 45–78, 169–170.

17. An alternative hypothesis asserts the opposite conclusion, that telegraphy could play a beneficial role by solving problems before they had time to fester and worsen. Sir John Pender, an entrepreneur involved in the telegraph cable industry, asserted, "No time was allowed for the growth of bad feeling or the nursing of a grievance. The cable nipped the evil of misunderstanding leading to war in the bud." Quoted in Headrick, *Invisible Weapon,* 74. While this hypothesis may contribute to an understanding of the underlying causes of some conflicts, I have not found it very useful for explaining the immediate origins of crises.

18. In fact, the United States had less cause for complaint in 1812 than in previous years, given British efforts to resolve sources of tension and a diminution in the number of American merchant ships seized under the Orders in Council. See Henry Adams, *History of the United States during the Administrations of Jefferson and Madison* (1891–1896; New York: Scribner, 1962), 6: 225; Ronald L. Hatzenbuehler and Robert L. Ivie, *Congress Declares War: Rhetoric, Leadership, and Partisanship in the Early Republic* (Kent, Oh.: Kent State University Press, 1983), 115, Reginald Horsman, *The Causes of the War of 1812* (Philadelphia: University of Pennsylvania Press, 1962), 14; Bradford Perkins, *Prologue to War: England and the United States, 1805–1812* (Berkeley: University of California Press, 1961), 3.

19. Perkins, *Prologue to War,* 144, 142–149; Spencer C. Tucker and Frank T. Reuter, *Injured Honor: The Chesapeake-Leopard Affair, June 22, 1807* (Annapolis: Naval Institute, 1996), 128 and ch. 7.

20. Arthur S. Link, *Wilson: The Struggle for Neutrality, 1914–1915* (Princeton: Princeton University Press, 1960), 372–379; Count Bernstorff, *My Three Years in America* (London: Skeffington, [1920]), 118–121; Gordon A. Craig, *Germany, 1866–1945* (New York: Oxford University Press, 1978), 369–370. Lansing to Bernstorff, 18 Dec. 1915, 763.72/2327–1/2A, and 20 Dec. 1915, 763.72/2328–1/2, reel 25, M367, NARA.

21. By July 1915 President Wilson hoped that the crisis period had passed. Additional sinkings (most notably the *Arabic* in August, the *Ancona* in November, and the *Persia* shortly thereafter) reopened the issue. Ernest R. May, *The World War and American Isolation, 1914–1917* (Cambridge, Mass.: Harvard University Press, 1959), 161–165.

22. Bernstorff to Bethmann Hollweg, 17 May 1915, in Bernstorff, *Three Years in America,* 26. Count Johann von Bernstorff, *Memoirs of Count Bernstorff* (New York: Random House, 1936), 141–142.

23. The U.S. government allowed Bernstorff to send coded messages over U.S. cables for the purpose of resolving the *Lusitania* dispute. Nonetheless, delays continued to hinder German diplomacy. Some of these could be attributed to British interference; Gerard to Lansing, 18 Dec. 1915, 763.72/2311, M367, reel 25, NARA. Other delays resulted from unwieldy arrangements that allowed Germany to send telegrams via the foreign ministries of the United States and Sweden. Bernstorff later

warned his government to use wireless when speed was of the essence because "cablegrams regularly require several days." Bernstorff to Bethmann Hollweg, #239, 27 Jan. 1917, in *The Papers of Woodrow Wilson,* ed. Arthur S. Link (Princeton: Princeton University Press, 1983), 41: 50, 52. See also Bernstorff, *Memoirs,* 142.

24. Robert Lansing, the U.S. secretary of state, acknowledged that "The protracted delay in the settlement of this controversy [is] due to unavoidable causes." Lansing to Bernstorff, 20 Dec. 1915, 763.72/2328–1/2, M367, reel 25, NARA. See also Bernstorff to Lansing, 29 Dec. 1915, 763.72/2338–1/2, M367, reel 25, NARA.

25. May, *World War and American Isolation,* 167.

26. Albert Sorel, "La diplomatie et le progrès," in Sorel, *Essais d'histoire et de critique* (Paris: Plon, 1883), 288–289. Information about anti-French feeling in Germany appears to have contributed to the change in French policy. A French investigation into German opinion during the 1840 crisis is available at Mémoires et Documents, Wurtemberg, 9, 341–379, MAE, Paris.

27. Quotation from [Gramont] to Benedetti, telegram, 12 July 1870, 228, Correspondance politique, Prusse, 379, MAE, Paris. See also Gramont to Benedetti, telegram, 10 July 1870, 130, and Benedetti to Gramont, 11 July 1870, 202–203, ibid.; Pierre Granet, *L'évolution des méthodes diplomatiques* (Paris: Rousseau, 1939), 85–90. On France in 1840, see Henry-Thierry Deschamps, *La Belgique devant la France de juillet: L'opinion et l'attitude françaises de 1839 à 1848* (Paris: Société d'Édition "Les Belles Lettres," 1956), 86–88; Pierre Renouvin, *Histoire des relations internationales: Le XIXe siècle: De 1815 à 1871* (1954; Paris: Hachette, 1994), 2: 432.

28. Likewise, an American historian argued that the availability of time for public reflection helps explain why the United States avoided war with Britain during the Oregon dispute of the mid-1840s, whereas the rush of events helped propel the United States into a war with Mexico at about the same time. Frederick Merk, *The Oregon Question: Essays in Anglo-American Diplomacy and Politics* (Cambridge, Mass.: Harvard University Press, 1967), 394.

29. May, *American Imperialism,* xx. While there is some merit to the notion that the media shape public attitudes and set the foreign policy agenda, political leaders also exert considerable influence on what the media report. See Johanna Neuman, *Lights, Camera, War: Is Media Technology Driving International Politics?* (New York: St. Martin's, 1996); Philip Seib, *Headline Diplomacy: How News Coverage Affects Foreign Policy* (Westport, Conn.: Praeger, 1997).

30. James A. Field Jr., "American Imperialism: The Worst Chapter in Almost Any Book," *American Historical Review* 83, no. 3 (June 1978), 663. "The Intellectual Effects of Electricity," *Spectator,* 9 Nov. 1889, 632.

31. In fact, telegraphy was not conducive to conveying rich detail because of the high expense involved—but scarcity of information can lead to an excitement of its own since it encourages use of the imagination.

32. No telegraph line connected Vienna and Constantinople. As a result, the news of the Sinope "massacre" did not reach Vienna until 10 December. The tele-

graph then carried this information to Paris on 11 December and London the next day. *Times* (London), 13 Dec. 1853, 7, 8.

33. Banquet speech, 15 Apr. 1864, in *Europe and America: Report on the Proceeding at an Inauguration Banquet Given by Mr. Cyrus Field* (London: Wm. Brown, 1864), box 142 (Cyrus Field papers), Collection #205 (Western Union Telegraph Company), 1993 Addendum, NMAH. Cyrus Field, a proponent of the Atlantic cable, heard Adams's statement and countered, "if we only had the telegraph in constant work between Britain and America, he would receive a short message, instead of a long despatch, and would reply to it with equal brevity." As a result, Field asserted, "instead of being hampered with business all day long, he would have plenty of time to bestow upon his numerous friends." Ibid.

34. Testimony by A. Buchanan, in *British Parliamentary Papers, First Report from the Select Committee on Diplomatic and Consular Services* (ordered by the House of Commons to be printed, 18 May 1871), 7: 13. See Zara S. Steiner, *Present Problems of the Foreign Service* (Princeton: Center of International Studies, 1961), 2–6.

35. Robert Bacon, 29 May 1907, vol. 6, E726 (Circulars of the Department of State), RG 59, NARA. Speech by Otto von Bismarck, *Verhandlungen des Reichstags* 79 (4 Dec. 1884), 198–199; M. S. Anderson, *The Rise of Modern Diplomacy, 1450–1919* (London: Longman, 1993), 104–105; Coates and Finn, *Retrospective Technology Assessment,* 91–93; Zara Steiner, "Introduction," in *The Times Survey of Foreign Ministries of the World,* ed. Steiner (London: Times Books, 1982), 13, 24.

36. C. W. de Kiewiet, *The Imperial Factor in South Africa: A Study in Politics and Economics* (Cambridge: Cambridge University Press, 1937), 293. Studies of communications volume during international crises indicate that greater stress produces heavier diplomatic traffic, and that sending telegrams may have served as a technique for stress management. Ole R. Holsti, *Crisis, Escalation, War* (Montreal: McGill-Queen's University Press, 1972), 83–95. Likewise, research in social psychology indicates that stress may heighten the desire for control and promote overcentralization. Dietrich Dörner, *The Logic of Failure: Why Things Go Wrong and What We Can Do to Make Them Right* (New York: Metropolitan Books, 1996), 162.

37. "Activity is certainly necessary and useful, but Lord Cromer wrote in his classic book: 'The masterpieces of statesman's art are, for the most part, not acts, but abstinences from action.'" Bernstorff, *Memoirs,* 285. Prince Bernard von Bülow quoted a Habsburg diplomat's proverb: "Patience is genius." *Memoirs of Prince von Bülow* (Boston: Little, Brown, 1932), 4: 13–14.

38. *Times* (London), 15 Nov. 1851, 4. The British government sent the first telegram from London to Vienna on 24 April 1852. Francis W. H. Cavendish, *Society, Politics and Diplomacy, 1820–1864, Passages from the Journal of Francis W. H. Cavendish* (London: T. Fisher Unwin, 1913), 232.

39. Richard Smoke, "The Crimean War," in *Avoiding War: Problems of Crisis Management,* ed. Alexander L. George (Boulder: Westview, 1991), 37.

40. Jerrold M. Post, "The Impact of Crisis-Induced Stress on Policy Makers," in

Avoiding War, ed. George, 474; Holsti, *Crisis, Escalation, War,* 11–13, 15; Richard Ned Lebow, *Between Peace and War: The Nature of International Crisis* (Baltimore: Johns Hopkins University Press, 1981), 107–111; Anne Edland and Ola Svenson, "Judgment and Decision Making under Time Pressure," in *Time Pressure and Stress in Human Judgment and Decision Making,* ed. Ola Svenson and A. John Maule (New York: Plenum, 1993), 36–37; Janice R. Kelly et al., "The Effects of Time Pressure and Task Difference on Influence Modes and Accuracy in Decision-Making Groups," *Personality and Social Psychology Bulletin* 23, no. 1 (Jan. 1997), 10–22; Eraldo de Grada et al., "Motivated Cognition and Group Interaction: Need for Closure Affects the Contents and Processes of Collective Negotiations," *Journal of Experimental Social Psychology* 35, no. 4 (July 1999), 346–365.

41. Irving L. Janis and Leon Mann, *Decision Making: A Psychological Analysis of Conflict, Choice, and Commitment* (New York: Free Press, 1977), 50–51, 59–62; Holsti, *Crisis, Escalation, War,* 134–138.

42. Janis and Mann, *Decision Making,* 50–51, 59–62. Post, "Impact of Crisis-Induced Stress," 475; Holsti, *Crisis, Escalation, War,* 13.

43. Clarendon to Stratford, 2 Aug. 1853, *House of Commons Parliamentary Papers,* 1854, vol. 71, pt. 2, 27; Allan Cunningham, *Eastern Questions in the Nineteenth Century,* ed. Edward Ingram (London: Frank Cass, 1993), vol. 2, 209–210, 212–213; Ann Pottinger Saab, *The Origins of the Crimean Alliance* (Charlottesville: University Press of Virginia, 1977), 74; Stratford Canning to Lord Clarendon, #25, 11 Aug. 1853, in Stanley Lane-Poole, *The Life of the Right Honourable Stratford Canning, Viscount Stratford de Redcliffe* (London: Longmans, Green, 1888), vol. 2, 291.

44. Lord John Russell to the Earl of Aberdeen, 23 Sept. 1853, in Walpole, *Life of Russell,* 2: 191. Clarendon to Russell, 24 Sept. 1853, in J. B. Conacher, *The Aberdeen Coalition, 1852–1855* (Cambridge: Cambridge University Press, 1968), 183. The French ambassador to Britain conveyed Clarendon's sense of urgency to Paris. See Walewski to French foreign minister, #35, 17 Sept. 1853, #38, 20 Sept. 1853, and #40, 23 Sept. 1853, Correspondance Politique, Angleterre, 691, MAE, Paris.

45. Russell to Clarendon, 4 Oct. 1853, in Muriel E. Chamberlain, *Lord Aberdeen: A Political Biography* (London: Longman, 1983), 487. Harold Temperley, *England and the Near East: The Crimea* (London: Longmans, Green, 1936), 352–354. Earl of Clarendon to Lord Cowley, #108, and Earl of Clarendon to Lord Stratford de Redcliffe, #109, both 23 Sept. 1853, and enclosure, in *House of Commons Parliamentary Papers,* 1854, vol. 71 (Eastern Papers, pt. 2), 114–116. R. W. Seton-Watson, *Britain in Europe, 1789–1914: A Survey of Foreign Policy* (1937; New York: Howard Fertig, 1968), 312; Lane-Poole, *Life of Canning,* 2: 307; Conacher, *Aberdeen Coalition,* 188–189. Stratford initially refused to obey this order, thereby delaying the arrival of the allied fleets in Constantinople. When he finally implemented the instruction, the crisis worsened considerably.

46. Hammond to Malmesbury, 17 Sept. 1858, in Raymond A. Jones, *The British Diplomatic Service, 1815–1914* (Gerrards Cross, U.K.: Colin Smythe, 1983), 123. Davison, *Unforgiving Minute,* 59–60. In 1890 Britain's foreign secretary commented

upon the insufficient information that telegrams provided. Jones, *British Diplomatic Service,* 125.

47. "Intellectual Effects of Electricity," *Spectator,* 9 Nov. 1889. See also Pradier-Fodéré, *Cours de droit diplomatique,* 1: viii, 214; 2: 338.

48. Coates and Finn, *Retrospective Technology Assessment,* 90.

49. Baron J. de Szilassy, *Traité pratique de diplomatie moderne* (Paris: Payot, 1928), 168. Joseph C. Grew, *Turbulent Era: A Diplomatic Record of Forty Years, 1904–1945,* ed. Walter Johnson (Boston: Houghton Mifflin, 1952), 221–222. *Verhandlungen des Reichstags: Stenographische Berichte* 79 (4 Dec. 1884), 199. Interview with Edmund Hammond, 14 Mar. 1870, in *British Parliamentary Papers, Report from the Select Committee on Diplomatic and Consular Services,* 1870, 7: 29.

50. Post, "Impact of Crisis-Induced Stress," 488–489; Alexander George, "Findings and Recommendations," in *Avoiding War,* ed. George, 559–560; Roy Lubit and Bruce Russett, "The Effects of Drugs on Decision-Making," *Journal of Conflict Resolution* 28, no. 1 (Mar. 1984), 85–102.

51. Grew, *Turbulent Era,* 316. Haniel to Baron Langwerth, 3 May 1919, R2589, AA-PA.

52. *Verhandlungen des Reichstags: Stenographische Berichte* 79 (4 Dec. 1884), 199. Testimony of Morier, 23 June 1870, in *British Parliamentary Papers, Report from the Select Committee,* 1870, 7: 325. Interview with H. G. Elliot, 24 Mar. 1870, ibid., 7: 74.

53. Elihu Root, confidential circular, 16 Jan. 1909, file no. 17530, M862, reel 997, NARA. In response to Root's circular, the U.S. ambassador to Japan informed the State Department "that during the past year and a half, messages of great length have been received and sent, and if the work of deciphering and enciphering had been confined to the Chief of Mission and the First Secretary, great delay and unusual labor would have resulted." The State Department agreed to grant an exemption to the Tokyo embassy. O'Brien to secretary of state, 19 Feb. 1909, and Knox to O'Brien, 24 Mar. 1909, file no. 16682, M862, reel 971, NARA.

54. One historian asserts, "Foreign employees had the run of the [U.S.] embassies, and it would have been little trouble for them to get hold of the books." Kahn, *Codebreakers,* 490.

55. Robert V. Kubicek, *The Administration of Imperialism: Joseph Chamberlain at the Colonial Office* (Durham: Duke University Press, 1969), 27–28.

56. Hammond to Cowley, Private, 14 Mar. 1856, FO 519/185, Cowley Papers, PRO. Notice by Chief Clerk, 30 Nov. 1906, 119.2, 1910–29, RG 59, NARA.

57. Roon Lübeck and W. Rau, 22 Nov. 1916, Washington Botschaft 724, AA-PA. The French Foreign Ministry suffered from both the delay and crossing of unnumbered telegrams during the diplomatic crisis that preceded the Franco-Prussian War. See Benedetti to Gramont, 10 and 11 July 1870, Correspondance politique, Prusse, 379, MAE, Paris.

58. Anderson, *Rise of Modern Diplomacy,* 118. Matthieu, circular to German diplomatic mission and consular officials, 18 Aug. 1913, R138281, AA-PA. The German navy decreed on 24 April 1907 that incoming ciphered telegrams from abroad

should be numbered. Letter to the chief of the Admiral Staff, 17 Dec. 1907, RM 5/765, BA-MA.

59. Alvey A. Adee, 14 Sept. 1900, vol. 5, E726, RG 59, NARA. Bryan to U.S. ambassador to France and others, 31 Aug. 1914, 119.2/110b, and Lansing to U.S. ambassador to Constantinople, circular, 5 Nov. 1914, 119.2/123a, 1910–29, RG 59, NARA.

60. This was particularly true at the U.S. State Department. Jerry Israel, "A Diplomatic Machine: Scientific Management in the Department of State, 1906–1924," in *Building the Organizational Society,* ed. Israel (New York: Free Press, 1972), 193.

61. H. J. Bruce, *Silken Dalliance* (London: Constable, 1947), 102.

62. James Joll, *The Origins of the First World War* (London: Longman, 1989), 29. Holsti, *Crisis, Escalation, War,* 86, 83–85, 92–95. From 23 July until 4 August 1914, the German foreign ministry received 111 telegrams from its embassy in London. During the previous five years it had received an average of 225 telegrams per year from the London embassy. I. A. Eingangsregister, Eingänge: Missione und Konsulate, AA-PA.

63. Telegram from Wollmann, 8 Sept. 1914, and Langwerth memorandum, 11 Sept. 1914, R2437, AA-PA. Facing a similar problem, Britain's Royal Navy reprimanded officials who sent unimportant telegrams marked "priority." It noted that "the use of 'priority' telegrams is restricted to matters of importance as well as urgency." 29 Dec. 1916, index heading 93, ADM 12/1567B, PRO.

64. Maurice Paléologue, *An Ambassador's Memoirs* (London: Hutchinson, 1923), 1: 37. Herrick to Bryan, 5 Aug. 1914, 119.2/85, 1910–29, RG 59, NARA. Charles S. Wilson to William Jennings Bryan, 6 Aug. 1914, 763.72/221, M367, reel 12, NARA. John Tilley and Stephen Gaselee, *The Foreign Office* (London: Putnam's, 1933), 173, 298. The Royal Navy also expressed concern over the failing eyesight of coding officers. Foreign Yards, 16 Feb. 1916, index heading 93, ADM 12/1567A, PRO.

65. Robert Lansing to Woodrow Wilson, 30 Jan. 1917, M743 (Personal and Confidential Letters from Secretary of State Lansing to President Wilson, 1915–1918), NARA; "Embassy Staff Disrupted," *New York Times,* 6 Jan. 1917, 3. Christopher Andrew, *For the President's Eyes Only: Secret Intelligence and the American Presidency from Washington to Bush* (New York: Harper Collins, 1995), 41; Herbert O. Yardley, *The American Black Chamber* (New York: Blue Ribbon Books, 1931), 21–23.

66. Matthieu, 24 Aug. 1914, R2437, AA-PA.

67. Holsti, *Crisis, Escalation, War,* 108.

68. Köhler to the state secretary of the Office of the Imperial Navy, 24 Dec. 1912, 127ff; Herringen to the state secretary of the Imperial Post Office (and following circular), 10 Feb. 1913, 131ff; and I. A., to the imperial chancellor, June 1913, 145ff, RM5/991, BA-MA. Germany was not alone in such contingency planning; Britain's Foreign Office prepared automatic telegraphic instructions to be used in the event of war. J. D. Gregory, *On the Edge of Diplomacy: Rambles and Reflections, 1902–1928* (London: Hutchinson, 1928?), 68.

69. Joll, *Origins of the First World War,* 27; Holsti, *Crisis, Escalation, War,* 104–118.

German officials alleged that France purposely jumbled two diplomatic telegrams sent to Paris on 3 August 1914. The claim of intentional garbling is highly unlikely, but, at least one (and perhaps both) of the telegrams was apparently garbled. Schoen to Auswärtiges Amt, 3 Aug. 1914, in *Outbreak of the World War: German Documents Collected by Karl Kautsky,* ed. Karl Kautsky, Max Montgelas, and Walther Schücking (New York: Oxford University Press, 1924), nos. 716, 776, 809, 521, 553, 569; Albertini, *Origins of the War,* 3: 214.

70. M. B. Hayne, *The French Foreign Office and the Origins of the First World War, 1898–1914* (Oxford: Clarendon, 1993), 279. Jules Laroche, *Au Quai d'Orsay avec Briand et Poincaré, 1913–1926* (Paris: Hachette, 1957), 20. Lebow, *Between Peace and War,* 129; Holsti, *Crisis, Escalation, War,* 115–116.

71. Friedrich von Prittwitz und Gaffron, *Zwischen Petersburg und Washington: Ein Diplomatenleben* (Munich: Isar Verlag, 1952), 69. Albertini, *Origins of the War,* 2: 610.

72. Das K. Haupttelegraphenamt an Auswärtiges Amt Chiffrierbureau, 2 Aug. 1914, telegrams of 7:40 P.M. and 10:00 P.M., T120, reel 1820 (Bd. 12), NARA. Exchange between E. Goschen and E. Grey, 2 Aug. 1914, in *British Documents on the Origins of the War: 1898–1914,* ed. G. P. Gooch and Harold Temperley, vol. 11 (London: HMSO, 1926), 273; E. Goschen to secretary of state [Jagow], 2 Aug. 1914, R19877, PA-AA.

73. Pradier-Fodéré, *Cours de droit diplomatique,* 2: 338.

74. Joll, *Origins of the First World War,* 173, 182–184. The U.S. ambassador to Berlin asserted: "The German lower classes . . . had no wish for war. In 1914 they had no time to protest. They were hurried into war before they knew what was happening." James W. Gerard, *My First Eighty-Three Years in America* (Garden City, N.Y.: Doubleday, 1951), 255.

75. Page to Bryan, #284, 6 Aug. 1914, 763.72/189, M367, reel 12, NARA.

5. Diplomatic Time

1. Any attempt to generalize about diplomatic culture across countries and decades runs the risk of eliding important distinctions while presenting a static, ahistorical picture. Nonetheless, such generalizations are useful in this case because of the striking conservatism of foreign policy institutions and the existence of a common diplomatic culture in the Western world before the First World War. See F. R. Bridge and Roger Bullen, *The Great Powers and the European States System, 1815–1914* (London: Longman, 1991), 19. In 1914, for example, the ambassadors to the Court of St. James from Germany, Austria-Hungary, and Russia were bound to one another, and to the British royal house, by family ties. Harry F. Young, *Prince Lichnowsky and the Great War* (Athens: University of Georgia Press, 1977), 49–50.

2. See, e.g., William D. Godsey Jr., *Aristocratic Redoubt: The Austro-Hungarian Foreign Office on the Eve of the First World War* (West Lafayette, Ind.: Purdue University Press, 1999), ch. 1; Helen Dittmer, "The Russian Foreign Ministry under Nicholas II: 1894–1914" (Ph.D. diss., University of Chicago, 1977), 88–89, 103;

R. J. B. Bosworth, *Italy, the Least of the Great Powers: Italian Foreign Policy before the First World War* (London: Cambridge University Press, 1979), 98, 102–103, 115; 595.

3. See, e.g., Lamar Cecil, *The German Diplomatic Service, 1871–1914* (Princeton: Princeton University Press, 1976), 66, 68, 323–327.

4. See, e.g., Raymond A. Jones, *The British Diplomatic Service, 1815–1914* (Gerrards Cross, U.K.: Colin Smythe, 1983), 139–146, 11–20. Bright quoted in Henry E. Mattox, *The Twilight of Amateur Diplomacy: The American Foreign Service and Its Senior Officers in the 1890s* (Kent: Kent State University Press, 1989), 115.

5. F. Charles-Roux, *Souvenirs diplomatiques d'un âge révolue* (Paris: Librairie Arthène Fatard, 1956), 7; Christophe Charle, "Noblesse et élites en France au début du XXe siècle," in *Les noblesses européennes au XIXe siècle,* Collection de l'École Française de Rome (Paris: Boccard, 1988), 410, 417. Christopher Andrew, *Théophile Delcassé and the Making of the Entente Cordiale: A Reappraisal of French Foreign Policy, 1898–1905* (London: Macmillan, 1968), 75; M. B. Hayne, *The French Foreign Office and the Origins of the First World War, 1898–1914* (Oxford: Clarendon, 1993), 24–25, 82.

6. Marcel Proust, *In Search of Lost Time: Within a Budding Grove* (London: Chatto and Windus, 1992), 2: 5. Hayne, *French Foreign Office,* 8; Adeline Daumard, "Noblesse et aristocratie en France au XIXe siècle," in *Les noblesses européennes au XIXe siècle,* 100; Dominic Lieven, *The Aristocracy in Europe, 1815–1914* (London: Macmillan, 1992), 187.

7. Hayne, *French Foreign Office,* 8, 306. John F. V. Keiger, *France and the Origins of the First World War* (London: Macmillan, 1983), 25; Paul Gordon Lauren, *Diplomats and Bureaucrats: The First Institutional Responses to Twentieth-Century Diplomacy in France and Germany* (Stanford: Hoover Institution Press, 1976), 27. A. J. P. Taylor, *The Struggle for Mastery in Europe, 1848–1918* (London: Oxford University Press, 1971), xxiii.

8. Eric Mension-Rigau, *Aristocrates et grands bourgeois: Éducation, traditions, valeurs* (Paris: Plon, 1994), ch. 6, 328, 345. Lieven, *Aristocracy in Europe,* 152–157. Central Europeans shared this infatuation with the style of the English gentleman. Young, *Prince Lichnowsky,* 25, 45. The Austro-Hungarian elite believed that the correct use of English slang provided evidence of nobility. Arthur J. May, *The Hapsburg Monarchy, 1867–1914* (Cambridge, Mass.: Harvard University Press, 1968), 162.

9. Archives Nationales, 427 (Archives Louis de Robien), Archives Privées, carton 5, 19–20, written 1943–1949. Talbot Imlay shared this material with me.

10. Warren Frederick Ilchman, *Professional Diplomacy in the United States, 1779–1939: A Study in Administrative History* (Chicago: University of Chicago Press, 1961), 74–77, 141; Mattox, *Twilight of Amateur Diplomacy,* 103–105, 42–43; David F. Long, *Gold Braid and Foreign Relations: Diplomatic Activities of U.S. Naval Officers, 1798–1883* (Annapolis: Naval Institute Press, 1988), 8.

11. Louis Kruh, "Stimson, the Black Chamber, and the 'Gentlemen's Mail' Quote," *Cryptologia* 12, no. 2 (Apr. 1988), 81; David Kahn, *The Codebreakers: The Story of Secret Writing* (New York: Scribner, 1997), 360.

12. F. P. Dunne, "Mr. Dooley on Diplomacy," *American Magazine* 66, no. 2 (June 1908), 107–111. Norman B. Ferris, "Lincoln and the Diplomatic Patronage," paper presented at the annual conference of the Society for Historians of American Foreign Relations, Princeton University, 25 June 1999, 8; Richard Hume Werking, *The Master Architects: Building the United States Foreign Service, 1890–1913* (Lexington: University of Kentucky Press, 1977), 14, 122.

13. Martin Weil, *A Pretty Good Club: The Founding Fathers of the U.S. Foreign Service* (New York: Norton, 1978), 46. Joseph McCarthy, "Speech at Wheeling, West Virginia," in *A History of Our Time: Readings on Postwar America,* ed. William H. Chafe and Harvard Sitkoff (New York: Oxford University Press, 1983), 61.

14. Cecil, *German Diplomatic Service,* 64, 63. Hayne, *French Foreign Office,* 14–15, 61, 151–152. Zara S. Steiner, *The Foreign Office and Foreign Policy, 1898–1914* (London: Ashfield, 1986), 184. D. C. M. Platt, *The Cinderella Service: British Consuls Since 1825* (London: Longman, 1971), ix, 1–4.

15. Steiner, *Foreign Office and Foreign Policy,* 185n1; Lamar Cecil, "Der diplomatische Dienst im kaiserlichen Deutschland," in *Das Diplomatische Korps, 1871–1945,* ed. Klaus Schwabe (Boppard am Rhein: Harald Boldt, 1985), 33; Jerry Israel, "A Diplomatic Machine: Scientific Management in the Department of State, 1906–1924," in *Building the Organizational Society,* ed. Israel (New York: Free Press, 1972), 187–190.

16. See Weil, *Pretty Good Club,* 47, 67–68; Mattox, *Twilight of Amateur Diplomacy,* 118; Cecil, *German Diplomatic Service,* 97–103; Zara S. Steiner, *Britain and the Origins of the First World War* (London: Macmillan, 1977), 174; Godsey, *Aristocratic Redoubt,* 86–89.

17. "The Need of a Trained Diplomatic Service," *New York Times,* 29 Apr. 1900.

18. Hayne, *French Foreign Office,* 8. Cecil, *German Diplomatic Service,* 39, 323–328. Lauren, *Diplomats and Bureaucrats,* 138–141. Enrico Serra, "Italy: The Ministry for Foreign Affairs," in Zara Steiner, ed., *The Times Survey of Foreign Ministries of the World* (London: Times Books, 1982), 298, 307, 308. Weil, *Pretty Good Club,* 46–47; Ilchman, *Professional Diplomacy,* 167–169. Baron Wilhelm von Schoen, *The Memoirs of an Ambassador: A Contribution to the Political History of Modern Times* (London: George Allen and Unwin, 1922), 121.

19. John Tilley and Stephen Gaselee, *The Foreign Office* (London: Putnam's, 1933), 195.

20. Thorstein Veblen, *The Theory of the Leisure Class* (1899; New York: Penguin, 1994), 43. Gordon W. Wood, *The Radicalism of the American Revolution* (New York: Vintage, 1991), 33, 36, 276. May, *Hapsburg Monarchy,* 162. The increasing popularity of sports among the "lower orders" produced an emphasis upon amateur athletics among the leisure class. Eligibility rules often excluded from competition anyone who had ever performed wage labor. Such barriers, based upon social snobbery, became enshrined in the Olympic movement. Allen Guttmann, *The Olympics: A History of the Modern Games* (Urbana: University of Illinois Press, 1992), 3–4, 12.

21. Edmund S. Morgan, *American Slavery, American Freedom: The Ordeal of Colo-*

nial Virginia (New York: Norton, 1975), 63. On conspicuous leisure by servants, see Veblen, *Theory of the Leisure Class,* 54–64. On foreign ministries as originating in the ruler's household, see Steiner, ed., *Times Survey,* 11. According to a British diplomat who received his first overseas posting in 1919, "When I joined [the diplomatic service] it was regarded as part of the King's Household and not really part of the Civil Service at all." David Kelly, *The Ruling Few or The Human Background to Diplomacy* (London: Hollis and Carter, 1953), 367.

22. Veblen, *Theory of the Leisure Class,* 63. See Joseph S. Roucek, "The Sociology of the Diplomat," *Social Science* 14, no. 4 (Oct. 1939), 370, 371.

23. See Lieven, *Aristocracy in Europe,* 134; Crown Prince Rudolf and Karl Menger, *Der oesterreichische Adel und sein constitutioneller Beruf. Mahnruf an die aristokratische Jugend* (Munich: Adolf Ackermann, 1878), esp. 12–14; May, *Hapsburg Monarchy,* 161–164. On bourgeois attitudes, see David S. Landes, *Revolution in Time: Clocks and the Making of the Modern World* (Cambridge, Mass.: Harvard University Press, 1983), 227–230; Lewis Mumford, *Technics and Civilization* (New York: Harcourt, Brace, 1934), 16; Peter Bailey, *Leisure and Class in Victorian England: Rational Recreation and the Contest for Control, 1830–1885* (London: Routledge and Kegan Paul, 1978), 63–64.

24. David Cannadine, *The Decline and Fall of the British Aristocracy* (New Haven: Yale Univ. Press, 1990), 281. Gordon A. Craig, "British Foreign Office from Grey to Austen Chamberlain," in *The Diplomats, 1919–1939,* ed. Gordon A. Craig and Felix Gilbert (New York: Atheneum, 1963), 1: 25. Professor J. S. Phillimore quoted in Algernon Cecil, *British Foreign Secretaries, 1807–1916* (London: G. Bell, 1927), 229.

25. Hayne, *French Foreign Office,* 24.

26. The war led to major reforms. See Lauren, *Diplomats and Bureaucrats;* J. D. Gregory, *On the Edge of Diplomacy: Rambles and Reflections, 1902–1928* (London: Hutchinson, 1928?), 18; Rachel West, *The Department of State on the Eve of the First World War* (Athens: University of Georgia Press, 1978), 137–138.

27. Tilley and Gaselee, *Foreign Office,* 144–145. Maurice Baring, *The Puppet Show of Memory* (1932; London: Cassell, 1987), 179.

28. Heinz Günther Sasse, "Zur Geschichte des Auswärtigen Amts," *100 Jahre Auswärtiges Amt, 1870–1970* (Bonn: Auswärtiges Amt, 1970), 28, 30; Moritz Busch, *Our Chancellor; Sketches for a Historical Picture,* trans. William Beatty-Kingston (New York: Scribner, 1884), 1: 237. Quotation from Lauren, *Diplomats and Bureaucrats,* 26.

29. Bosworth, *Italy,* 96. Teddy J. Uldricks, "The Tsarist and Soviet Ministry of Foreign Affairs," Steiner, ed., *Times Survey,* 522.

30. Delavaud to Delcassé, 17 July 1903, PA-AP, Delcassé, vol. 4, MAE, Paris; Hayne, *French Foreign Office,* 12, 15. Charles-Roux, *Souvenirs diplomatiques,* 92, 96. Steiner, *Foreign Office and Foreign Policy,* 229.

31. Andrew, *Théophile Delcassé,* 77; Charles-Roux, *Souvenirs diplomatiques,* 90–91; Keiger, *France and the Origins of the First World War,* 48. On when the French Foreign Ministry embraced new technologies, compare Richard D. Challener, "The

French Foreign Office: The Era of Philippe Berthelot," in *The Diplomats,* ed. Craig and Gilbert, 1: 60, with Jules Laroche, *Au Quai d'Orsay avec Briand et Poincaré, 1913–1926* (Paris: Hachette, 1957), 10–12.

32. H. J. Bruce, *Silken Dalliance* (London: Constable, 1947), 86. Valerie Cromwell and Zara S. Steiner, "The Foreign Office before 1914: A Study in Resistance," in *Studies in the Growth of Nineteenth-Century Government,* ed. Gillian Sutherland (Totowa, N.J.: Rowman and Littlefield, 1972), 182. Tilley and Gaselee, *Foreign Office,* 131–132.

33. Steiner, *Foreign Office and Foreign Policy,* 22, 183. Jones, *British Diplomatic Service,* 100. Gregory, *Edge of Diplomacy,* 56–57.

34. Lawrence S. Kaplan, "The Brahmin as Diplomat in Nineteenth Century America: Everett, Bancroft, Motley, Lowell," *Civil War History* 19, no. 1 (Mar. 1973), 5–28. Robert D. Schulzinger, *The Making of the Diplomatic Mind: The Training, Outlook, and Style of United States Foreign Service Officers, 1908–1931* (Middletown: Wesleyan University Press, 1975), 3, 19, 20, 21; Werking, *Master Architects,* 121–122; Weil, *Pretty Good Club,* 18–21.

35. See, e.g., Freiherr von Musulin, *Das Haus am Ballplatz: Erinnerungen eines österreich-ungarischen Diplomaten* (Munich: Verlag für Kulturpolitik, 1924), 37; Young, *Prince Lichnowsky,* 8–9; Count Bernstorff, *Memoirs of Count Bernstorff* (New York: Random House, 1936), 36, 54–55. Collier to Root, 26 Apr. 1907, in Werking, *Master Architects,* 121.

36. Grantzius to Speck von Sternburg, 31 Mar. 1903, Washington Botschaft #420, AA-PA. Akten betreffend der Sommeraufenthalt der Botschaft, Mar. 1913–Feb. 1916, Washington Botschaft #23, AA-PA. Friedrich von Prittwitz und Gaffron, *Zwischen Petersburg und Washington: Ein Diplomatenleben* (Munich: Isar, 1952), 35.

37. Lloyd C. Griscom, *Diplomatically Speaking* (Boston: Little, Brown, 1940), 69. Musulin, *Das Haus am Ballplatz,* 94–95; John MacMurray to Edna MacMurray, 22 Apr. 1909, John van Antwerp MacMurray Collection, box 15, Princeton University; Charles-Roux, *Souvenirs diplomatiques,* 19–20.

38. Baring, *Puppet Show of Memory,* 188. Bruce, *Silken Dalliance,* 96, 100. See diary entries for 1909, MacMurray Collection, box 15; Anton Graf Monts, *Erinnerungen und Gedanken des Botschafters Anton Graf Monts,* ed. Karl Friedrich Nowak and Friedrich Thimme (Berlin: Verlag für Kulturpolitik, 1932), 112.

39. Monts, *Erinnerungen und Gedanken,* 63; Bruce, *Silken Dalliance,* 101, 104; Emerich Csáky, *Vom Geachteten zum Geächteten: Erinnerungen des k. und k. Diplomaten und k. ungarischen Aussenministers,* ed. Eva-Marie Csáky (Vienna: Böhlau Verlag, 1992), 130; Röhl, *Kaiser and His Court,* 100; Godsey, *Aristocratic Redoubt,* 200. On the careers of homosexuals in the Austro-Hungarian diplomatic corps, see Godsey, *Aristocratic Redoubt,* 91–92. Nonetheless, responses were not uniform and one can find examples of both tolerance and homophobia during this time.

40. Keith Hamilton and Richard Langhorne, *The Practice of Diplomacy: Its Evolution, Theory and Administration* (London: Routledge, 1995), 130; Lieven, *Aristocracy in Europe,* 150–151; Elihu B. Washburne, *Recollections of a Minister to France:*

1869–1877 (New York: Scribner's, 1887), vol. 1; Monts, *Erinnerungen und Gedanken,* 112–113. On the division between work and leisure, see Steven M. Gelber, *Hobbies: Leisure and the Culture of Work in America* (New York: Columbia University Press, 1999), x, 1, 295–296; Michael Adas, *Machines as the Measure of Men: Science, Technology, and Ideologies of Western Dominance* (Ithaca: Cornell University Press, 1989), 242, 251.

41. Otto Pflanze, *Bismarck and the Development of Germany* (Princeton: Princeton University Press, 1990), 2: 364–367. Godsey, *Aristocratic Redoubt,* 200–201. Young, *Prince Lichnowsky,* 103.

42. Woodrow Wilson to John MacMurray, 21 Dec. 1905, MacMurray Collection, box 12. Aubrey Leo Kennedy, *Old Diplomacy and New, 1876–1922, From Salisbury to Lloyd-George* (New York: Appleton, 1923), 364–365. Count Johann Heinrich von Bernstorff, Germany's ambassador to Washington during Wilson's administration, believed Wilson biased against aristocrats. Bernstorff, *My Three Years in America* (London: Skeffington, 1920), 22.

43. Isabel V. Hull, "Kaiser Wilhelm II and the 'Liebenberg Circle,'" in *Kaiser Wilhelm II: New Intepretations, The Corfu Papers,* ed. John C. G. Röhl and Nicolaus Sombart (Cambridge: Cambridge University Press, 1982), 193, 194, 201, 210, 211; Röhl, *Kaiser and His Court,* 28; James D. Steakley, "Iconography of a Scandal: Political Cartoons and the Eulenburg Affair in Wilhelmin Germany," in *Hidden From History: Reclaiming the Gay and Lesbian Past,* ed. Martin Bauml Duberman et al. (New York: New American Library, 1989), 236–237, 246.

44. Steakley, "Iconography of a Scandal," 249, 253; Magnus Hirschfeld, *Von einst bis jetzt: Geschichte einer homosexuellen Bewegung, 1897–1922,* ed. Manfred Herzer and James Steakley (Berlin: Rosa Winkel, 1986), 87–92. Quotation from Maximilian Harden, "Prozess Eulenburg," *Die Zukunft* 15, no. 43 (25 July 1908), 133.

45. In 1918 rightists opposed to the policies of the foreign minister, Richard von Kühlmann, spread scandalous tales about his off-duty activities and eventually obtained his removal from office. David Blackbourn, *Fontana History of Germany, 1780–1918, The Long Nineteenth Century* (London: Fontana, 1997), 487–488; Richard von Kühlmann, *Erinnerungen* (Heidelberg: Verlag Lambert Schneider, 1948), 562–565; H. A. P., "Alldeutsche Sittenrichter," *Deutsche Zeitung* 23, #204 (23 Apr. 1918, Morgen Ausgabe), 1–2.

46. Raymond A. Jones, *Arthur Ponsonby: The Politics of Life* (London: Christopher Helm, 1989), 19, 71–73. Henry Adams, *The Education of Henry Adams: An Autobiography* (Boston: Houghton Mifflin, 1918), 210. [Arthur Ponsonby], "Diplomacy as a Profession," *National Review* (London), 35 (Mar. 1900), 103. "Arthur Ponsonby, the Diplomatic Service and the Foreign Office, 1900–1902," in Steiner, *Foreign Office and Foreign Policy,* app. 4.

47. Veblen, *Theory of the Leisure Class,* 42–43. Morgan, *American Slavery, American Freedom,* 83–84, 86, 72–73. Bernard Bailyn et al., *The Great Republic: A History of the American People* (Lexington, Mass.: Heath, 1992), 1: 42.

48. See, e.g., Michael Sanderson, *The Universities and British Industry, 1850–1970*

(London: Routledge and Kegan Paul, 1972), 4–8; Sidney Pollard, *Britain's Prime and Britain's Decline: The British Economy, 1870–1914* (London: Edward Arnold, 1989), 118–121, 182–183; Correlli Barnett, *The Audit of War: The Illusion and Reality of Britain as a Great Nation* (London: Macmillan, 1986), 216–220.

49. Kubicek, *Administration of Imperialism,* 31–33. Bosworth, *Italy,* 118. Holger H. Herwig, *"Luxury" Fleet: The Imperial German Navy 1888–1918* (London: Ashfield, 1987), 127–130. H. Sasse, "Von Equipage und Automobilen des Auswärtigen Amts," *Vereinigung deutscher Auslandsbeamten e. V. Nachrichtenblatt* 20, no. 10 (Oct. 1957), 145–148.

50. Dennis Showalter, "Soldiers into Postmasters: The Electric Telegraph as an Instrument of Command in the Prussian Army," *Military Affairs* 37 (Apr. 1973), 48–52. Frank Thomas, *Telefonieren in Deutschland: Organisatorische, technische und räumliche Entwicklung eines grosstechnischen Systems* (Frankfurt: Campus, 1995), 180–182, 186–187. Tschirschky to German Foreign Ministry, #105, 26 July 1914, T120, reel 1818 (Bd. 5), NARA.

51. The standard operating procedures of bureaucracies are based on resisting change; hence the adage, "Never do anything for the first time." James Q. Wilson, *Bureaucracy: What Government Agencies Do and Why They Do it* (New York: Basic Books, 1989), 221.

52. Steiner, *Foreign Office and Foreign Policy,* 16–18; Ray Jones, *The Nineteenth-Century Foreign Office: An Administrative History* (London: Weidenfeld and Nicolson, 1971), 64; Cecil, *German Diplomatic Service,* 26–39; Hampe, *Das Auswärtige Amt in der Ära Bismarck,* 57–60; Georges Dethan, "France: The Ministry of Foreign Affairs since the Nineteenth Century," in Steiner, ed., *Times Survey,* 205, 209; Uldricks, "Tsarist and Soviet Ministry of Foreign Affairs," 518; Serra, "Italy: The Ministry for Foreign Affairs," 298, 307.

53. Lauren, *Diplomats and Bureaucrats,* 64. Zara S. Steiner, *Present Problems of the Foreign Service* (Princeton: Center of International Studies, 1961), 2–6.

54. Seton Gordon, *Edward Grey of Fallodon and His Birds* (London: Country Life, 1937), 16–17; George Macaulay Trevelyan, *Grey of Fallodon: The Life and Letters of Sir Edward Grey* (Boston: Houghton Mifflin, 1937), 388. Hammond quoted in Daniel Headrick, *The Invisible Weapon: Telecommunications and International Politics, 1851–1945* (New York: Oxford University Press, 1991), 74. Waldo H. Heinrichs Jr., *American Ambassador: Joseph Grew and the Development of the United States Diplomatic Tradition* (New York: Garland, 1979), 13.

55. Friedrich von Holstein to Ida von Stülpnagel, 27 Mar. 1902, in *Historische Augenblicke: Deutsche Briefe des Zwanzigsten Jahrhunderts,* ed. Jürgen Moeller (Munich: Beck, 1988), 19.

56. Volker Aschoff, "Paul Schilling von Canstatt und die Geschichte des elektromagnetischen Telegraphen," *Abhandlungen und Berichte des Deutschen Museums* 44, no. 3; Keith Dawson, "Electromagnetic Telegraphy: Early Ideas, Proposals, and Apparatus," *History of Technology* 1 (1976).

57. "The Intellectual Effects of Electricity," *Spectator,* 9 Nov. 1889, 632. Lady

Gwendolen Cecil, *Life of Robert, Marquis of Salisbury* (London: Hodder and Stoughton, 1931), 3: 6–7; Lord David Cecil, *The Cecils of Hatfield House* (London: Constable, 1973), 239–240.

58. "Intellectual Effects of Electricity," *Spectator,* 9 Nov. 1889. Lieven, *Aristocracy in Europe,* 151, 152, 158, 160.

59. Edward Tenner, *Why Things Bite Back: Technology and The Revenge of Unintended Consequences* (New York: Knopf, 1996); Robert L. Heilbroner, "Do Machines Make History?" in *Does Technology Drive History: The Dilemma of Technological Determinism,* ed. Merritt Roe Smith and Leo Marx (Cambridge, Mass.: MIT Press, 1995), 54–65.

60. Christoph von Tiedemann, *Sechs Jahre Chef der Reichskanzlei unter dem Fürsten Bismarck* (Leipzig: Hirzel, 1910), 210, 312; Moritz Busch, *Bismarck: Some Secret Pages of His History* (New York: AMS Press, 1970), 2: 180, 188–189. Karl-Alexander Hampe, *Das Auswärtige Amt in der Ära Bismarck* (Bonn: Bouvier Verlag, 1995), 40.

61. Otto Pflanze, *Bismarck and the Development of Germany* (Princeton: Princeton University Press, 1990), 3: 303–307, 338, 351–355; A. J. P. Taylor, *Bismarck: The Man and the Statesman* (New York: Vintage, 1967), 234–236.

62. Viscount Grey of Fallodon, *Fly Fishing* (1899; London: Dent, 1930); Viscount Grey of Fallodon, *The Charm of Birds* (New York: Stokes, 1927). Keith Robbins, *Sir Edward Grey: A Biography of Lord Grey of Fallodon* (London: Cassell, 1971), xiv.

63. Quotation from Trevelyan, *Grey of Fallodon,* 189; Taylor, *Struggle for Mastery,* xxiii; John Wilson, *CB: A Life of Sir Henry Campbell-Bannerman* (London: Constable, 1973), 435, 439, 455–456; Cecil, *British Foreign Secretaries,* 319–320. Most Liberal peers had left for the Conservative party; Grey was one of the few remaining Liberals with a pedigree worthy of the highest post in the Foreign Office.

64. Cannadine, *Decline and Fall of the British Aristocracy,* 215. David Lloyd George, *War Memoirs of David Lloyd George, 1914–1915* (Boston: Little, Brown, 1933), 1: 88–89.

65. Viscount Grey of Fallodon, *Twenty-Five Years, 1892–1916* (New York: Stokes, 1925), 1: 302, 305. Lloyd George, *War Memoirs,* 1: 51; "Policy of Great Britain," London *Times,* 27 July 1914, 7; Harold Nicolson, *Sir Arthur Nicolson, Bart., First Lord Carnock: A Study in the Old Diplomacy* (London: Constable, 1930), 413.

66. Lichnowsky to German Foreign Ministry, 26 July 1914, #160, T120, reel 1818 (Bd. 5), NARA. Lichnowsky sought "to spare the German people a struggle, from which they have nothing to win and everything to lose." Lichnowsky to German Foreign Ministry, #161, 26 July 1914, T120, reel 1818 (Bd. 6), NARA.

67. See Jonathan Powis, *Aristocracy* (Oxford: Basil Blackwell, 1984), 94; F. M. L. Thompson, *English Landed Society in the Nineteenth Century* (London: Routledge and Kegan Paul, 1963), 1, 258; Cannadine, *Decline and Fall of the British Aristocracy,* 359, 370, 385; Blackbourn, *Fontana History of Germany, 1780–1918,* 119, 280.

68. Nineteenth-century observers invoked "the triad of railroad, steamship, telegraph" to represent industrial progress. Wolfgang Schivelbusch, *The Railway Jour-*

ney: *The Industrialization of Time and Space in the Nineteenth Century* (Berkeley: University of California Press, 1986), 194. On the importance of telegraphy and the railroad to the development of "large modern manufacturing or marketing enterprises," see Alfred D. Chandler Jr., *The Visible Hand: The Managerial Revolution in American Business* (Cambridge, Mass.: Harvard University Press, 1977), 79.

69. Davis Bowman, *Masters and Lords: Mid-19th-Century U.S. Planters and Prussian Junkers* (New York: Oxford University Press, 1993), 28; Veblen, *Theory of the Leisure Class,* 64–65; V. Wheeler-Holohan, *The History of the King's Messengers* (London: Grayson and Grayson, 1935), 247.

70. Godsey, *Aristocratic Redoubt,* 198. Adams, *Education,* 5.

71. Busch, *Our Chancellor,* 1: 169–170; Tibor Süle, *Preussische Bürokratiertradition: Zur Entwicklung von Verwaltung und Beamtenschaft in Deutschland, 1871–1918* (Göttingen: Vandenhoeck & Ruprecht, 1988), 36. Joseph Maria von Radowitz, *Aufzeichnungen und Erinnerungen aus dem Leben des Botschafters Joseph Maria von Radowitz,* ed. Hajo Holborn (Osnabrück: Biblio-Verlag, 1967), 1: 259. I have borrowed the notion of "task orientation" from E. P. Thompson, "Time, Work-Discipline, and Industrial Capitalism," *Past and Present,* no. 38 (Dec. 1967), 60.

72. Thompson, "Time, Work-Discipline, and Industrial Capitalism," 60; Mumford, *Technics and Civilization,* 16; Nancy F. Koehn, "Josiah Wedgwood and the First Industrial Revolution," in *Creating Modern Capitalism: How Entrepreneurs, Companies, and Countries Triumphed in Three Industrial Revolutions,* ed. Thomas K. McCraw (Cambridge, Mass.: Harvard University Press, 1997), 42–44.

73. Schoen to all consuls, 22 Jan. 1909, R138730, AA-PA. One historian argues that the link "between the clock and the telegraph" set the stage for "Taylorist doctrines of scientific management, with its strict calculus of temporal work efficiency." Graeme Davison, *The Unforgiving Minute: How Australia Learned to Tell the Time* (Melbourne: Oxford University Press, 1993), 56.

74. See Landes, *Revolution in Time,* 285–286; Vary T. Coates, Bernard Finn, et al., *A Retrospective Technology Assessment: Submarine Telegraphy: The Transatlantic Cable of 1866* (San Francisco: San Francisco Press, 1979), 182; Davison, *Unforgiving Minute,* 2, 4, 38, 40, 47, 52–56. In 1852 an Exeter newspaper observed, "The completion of the electric telegraph from the metropolis [London] to this city, has naturally brought under discussion the question of uniformity of time." See Clive N. Ponsford, *Time in Exeter* (Exeter: Headwell Vale, 1978), 13, 11. In the United States, the Western Union Telegraph Company established a commercial "Time Signal Service" that transmitted standard time over telegraph wires from the U.S. Naval Observatory in Washington, D.C., to the rest of the nation. WUTC, Collection #205, ser. 3, box 14, folders 1 and 3, NMAH.

75. Bruce, *Silken Dalliance,* 102. Frederic Rogers to Lady Rogers, Nov. 1869, in *Letters of Frederic Lord Blachford, Under-secretary of State for the Colonies,* ed. George Eden Marindin (London: J. Murray, 1896), 278–279.

76. Hammond to Malmesbury, 21 Sept. 1858, in Jones, *British Diplomatic Service,* 124. J. R. Pole quoted in Bowman, *Masters and Lords,* 29. The final quotation con-

cerns everyone, not just aristocrats; "Intellectual Effects of Electricity," *Spectator,* 9 Nov. 1889. *Little Arthur's History of England* (1835) declared: "The nobles of England are useful to the country. As they are rich enough to live without working for themselves and their families, they have time to be always ready when the King wants advice, or when there is a Parliament to make laws or when the King wishes to send messages to other kings." Quoted in Jones, *Ponsonby,* 71.

77. Schulzinger, *Making of the Diplomatic Mind,* 16. Hamilton Fish, Circular, 1 Nov. 1873, #47, vol. 1, E726 (Circulars of the Department of State), RG 59, NARA; Algernon Cecil, "The Foreign Office," in *The Cambridge History of British Foreign Policy, 1783–1919,* ed. A. W. Ward and G. P. Gooch, vol. 3 (Cambridge: Cambridge University Press, 1923), 590; French Foreign Ministry aide mémoire, draft, 6 July 1914, Personnel/Dossier Généraux (1706–1940)/206, MAE, Paris.

78. Norman Rich, *Friedrich von Holstein: Politics and Diplomacy in the Era of Bismarck and William II* (Cambridge: Cambridge University Press, 1965), 1: 75. Godsey, *Aristocratic Redoubt,* 179. Bosworth, *Italy,* 96. Cecil, *German Diplomatic Service,* 62.

79. Busch, *Our Chancellor,* 1: 237–239. Harold Nicolson, *Diplomacy* (London: Thornton Butterworth, 1939), 75. Gregory, *Edge of Diplomacy,* 14.

80. Gordon Craig, "Bismarck and His Ambassadors: The Problem of Discipline," in Craig, *War, Politics, and Diplomacy: Selected Essays* (New York: Praeger, 1966), 186.

81. Stratford Canning to his wife, 13 Sept. 1853, in Stanley Lane-Poole, *The Life of the Right Honourable Stratford Canning, Viscount Stratford de Redcliffe* (London: Longmans, Green, 1888), 2: 300.

82. Algernon Cecil, "Foreign Office," 586; Mary Adeline Anderson, "Edmund Hammond, Permanent Under Secretary of State for Foreign Affairs, 1854–1873" (Ph. D. diss., University of London, 1956), 82–83.

83. Jacques Baeyens, *Au Bout du Quai: Souvenirs d'un retraité des postes* (Paris: Fayard, 1975), 222.

84. Robert K. Massie, *Dreadnought: Britain, Germany, and the Coming of the Great War* (New York: Random House, 1991), 736. Churchill to David Lloyd George, 14 Sept. 1911, in Randolph S. Churchill, *Winston S. Churchill: Young Statesman, 1901–1914* (Boston: Houghton Mifflin, 1967), 2: 515–516. Jules Laroche, *Quinze ans à Rome avec Camille Barrère, 1898–1913* (Paris: Plon, 1948), 285.

85. Cecil, "Der diplomatische Dienst im kaiserlichen Deutschland," 29–30, 18. Harry Graf Kessler, *Gesichter und Zeiten: Erinnerungen, Völker und Vaterländer* (Berlin: S. Fischer, 1935), 234. On Bismarck's pursuit of Isabella Lorraine-Smith, see Taylor, *Bismarck,* 18; Lothar Gall, *Bismarck: Der weisse Revolutionär* (Frankfort: Ullstein, 1993), 39–40.

86. Statement by Bebel, *Verhandlungen des Reichstages* 144 (14 Feb. 1896), 955.

87. "On Vacation: An Unconventional Ambassador," *Sketch* 96, no. 1239 (25 Oct. 1916), 73; conversation between Alice White and Hans Bernstorff, 14 Nov. 1916, folder 0029, box 38, ser. IV, group no. 656, Frank L. Polk Papers, Yale University Manuscripts and Archives. The photo appears in Barbara Tuchmann, *The Zimmer-*

mann Telegram (New York: Macmillan, 1966), opposite 155. Gerard to Lansing, 3 Jan. 1917, 763.72/3107–1/2, M367, reel 31, NARA; Tuchmann, *Zimmermann Telegram,* 125–126. Ritter, *Sword and Scepter,* 4: 183–187; Kühlmann, *Erinnerungen,* 563–564. Evidence of Bernstorff's indiscreet lifestyle can be found in telephone conversations (which U.S. officials secretly transcribed) by and about him during the First World War. See, e.g., conversation of 22 Apr. 1916, 11:03 am (2), folder 0011, box 38, ser. IV, group no. 656, Polk Papers.

88. Schoen to all consuls, 22 Jan. 1909, R138730, AA-PA; unsigned [Kiderlen-Wächter], Verfügung, 12 Apr. 1911, R139522, AA-PA; Kiderlen-Wächter, Verfügung, 7 May 1912, R139523, AA-PA; von Heeringen to all Staatsministers, 14 May 1912, R139523, AA-PA.

89. Unsigned [Kiderlen-Wächter], Verfügung, 12 Apr. 1911, R139522, AA-PA. Wilhelm von Schoen to Hardenberg, 6 May 1909, R139522, AA-PA. Freiherr von Richthofen to all imperial bureaucrats in the consular service, 6 Dec. 1899, R139522, AA-PA. Schoen to all consuls, 22 Jan. 1909, R138730, AA-PA.

90. German consular officers sometimes could not afford to return to Germany for vacations. See dispatch with illegible signature (written from the consulate in Shanghai) to Bethmann Hollweg, 28 Aug. 1911, R139523, AA-PA, Bonn; Schoen to state secretary of the Imperial Treasury, 30 June 1910, R139522, AA-PA.

91. Kühlmann, *Erinnerungen,* 295. Cannadine, *Decline and Fall of the British Aristocracy,* 281; Griscom, *Diplomatically Speaking,* 51; Monts, *Erinnerungen und Gedanken,* 112. Leonore Davidoff, *The Best Circles: Society Etiquette and the Season* (London: Cresset, 1986), 30–31.

92. Ada v. Erdmann, *Nikolaj Karlovic Giers, russischer Aussenminister, 1882–1895* (Inaugural-Dissertation zur Erlangung der Doktorwürde, Friedrich-Wilhelms-Universität zu Berlin, 1935?), 11. J. A. S. Grenville, *Lord Salisbury and Foreign Policy: The Close of the Nineteenth Century* (London: Athlone, 1964), 62; Ernest R. May, *Imperial Democracy: The Emergence of America as a Great Power* (Chicago: Imprint Publications, 1991), 43; Gerald G. Eggert, *Richard Olney: Evolution of a Statesman* (University Park: Pennsylvania State University Press, 1974), 215.

93. Alex Hall, *Scandal, Sensation and Social Democracy* (Cambridge: Cambridge University Press, 1977); 162; Katharine Anne Lerman, *The Chancellor as Courtier: Bernhard von Bülow and the Governance of Germany, 1900–1909* (Cambridge: Cambridge University Press, 1990), 221–222; Cecil, *German Diplomatic Service,* 302–305; Lamar Cecil, *Wilhelm II: Emperor and Exile, 1900–1941* (Chapel Hill: University of North Carolina Press, 1996), 2: 133–134.

94. Le Sourd to Gramont, #94, 8 July 1870, 61–62, Correspondance Politique, Prusse, 379, MAE, Paris. Goeppert report, 12, 4 Apr. 1917, Berlin, R16919, AA-PA.

95. William M. Johnston, *The Austrian Mind: An Intellectual and Social History, 1848–1938* (Berkeley: University of California Press, 1972), 37, 41. Musulin, *Das Haus am Ballplatz,* 213, 214, 223. Josef Redlich, *Das politische Tagebuch Josef Redlichs, 1908–1919* (Graz: Hermann Böhlaus, 1953), 1: 234. *Mezhdunarodnye otnosheniya v epokhu imperializma* (Moscow: State Social Economic Press, 1931), 3rd ser., vol. 4, 15/

2 July 1914, #238, 284–285. Albertini suggests that the passivity of the Russian legation in Belgrade may have worsened the July crisis. Luigi Albertini, *The Origins of the War of 1914* (London: Oxford University Press, 1953), 2: 279.

96. For example, the vacation schedule for the secret chancellery of the German foreign ministry indicates that this office reached its nadir in personnel during the month of July. "Urlaubs Ubersicht der Beamten der Geheimen Kanzlei des Auswärtigen Amts für 1914," R139567, AA-PA, Bonn.

97. *Times* (London), 20 July 1914, 12; 22 July 1914, 11; Ross Gregory, *Walter Hines Page: Ambassador to the Court of St. James's* (Lexington: University Press of Kentucky, 1970), 49–50. Tucher an Hertling, 23 July 1914, in *Julikrise und Kriegsausbruch 1914,* ed. Imanuel Geiss (Hannover: Verlag für Literatur und Zeitgeschehen, 1963), vol. 1, no. 241, 312. Hayne, *French Foreign Office,* 292.

98. Bernstorff, *Three Years in America,* 30–32; "Policy of Great Britain," *Times* (London), 27 July 1914, 7; Edward Goschen, *The Diary of Edward Goschen, 1900–1914,* ed. Christopher H. D. Howard (London: Royal Historical Society, 1980), 40–41. Albertini, *Origins of the War,* 2: 282, 347–348. "Diplomacy at Work," *Times* (London), 27 July 1914, 8; Serge Sazonov, *Fateful Years: 1909–1916* (New York: Stokes, 1928), 152.

99. Godsey, *Aristocratic Redoubt,* 190. James Joll, *The Origins of the First World War* (London: Longman, 1989), 21; William Jannen Jr., *The Lions of July: Prelude to War, 1914* (Novato, Calif.: Presidio, 1996), 15–16; Samuel R. Williamson Jr., *Austria-Hungary and the Origins of the First World War* (London: Macmillan, 1991), 208, 214, 203, 206.

100. In a discussion with the Austro-Hungarian foreign minister following the ultimatum to Serbia, Russia's chargé d'affaires, Prince Kudascheff (Kudashev), rather than warning of the danger of war, noted that he had received no instructions and merely referred matters back to St. Petersburg. Tschirschky to German Foreign Ministry, #101, 24 July 1914, T120, reel 1818 (Bd. 4), NARA.

101. Williamson, *Austria-Hungary and the Origins of the War,* 195. Cecil, *Wilhelm II,* 2: 201; Isabel V. Hull, *The Entourage of Kaiser Wilhelm II, 1888–1918* (Cambridge: Cambridge University Press, 1982), 36–37.

102. Cecil, *Wilhelm II,* 2: 193–197, 201–203. Albertini, *Origins of the War,* 2: 436. Wilhelm II, *The Kaiser's Memoirs* (New York: Harper, 1922), 246–248. Joll, *Origins of the First World War,* 18, 22–23. Grand Admiral von Tirpitz, *My Memoirs* (New York: Dodd, Mead, 1919), 1: 317. Other contemporaries also believed the Kaiser's absence made a difference. See Countess Olga Leutrum, *Court and Diplomacy in Austria and Germany: What I Know* (Philadelphia: Lippincott, 1918), 163–164.

103. Albertini, *Origins of the War,* 2: 436–437. Cecil, *Wilhelm II,* 2: 416n48, 201–202. The Belgian ambassador to Berlin felt reassured when he heard that Wilhelm II and much of the diplomatic corps in Berlin were taking their vacations in July despite rumors of an impending crisis. Baron Beyens, *Deux années à Berlin, 1912–1914* (Paris: Plon, 1931), 226.

104. *Mezhdunarodnye otnosheniya v epokhu imperializma,* 3rd ser., vol. 4, 15/2 July

1914, #236, 283. Adolf Hitler later used similar tactics, timing international crises so that they fell on weekends. The junior officials who remained to staff the foreign ministries and embassies lacked the experience and authority to respond quickly to Hitler's unexpected maneuvers. Personal communication from Zachary Shore, 28 Nov. 1999; Joachim C. Fest, *Hitler* (New York: Harcourt Brace Jovanovich, 1974), 490, 497; Zach Shore, "Hitler, Intelligence and the Decision to Remilitarize the Rhine," *Journal of Contemporary History* 34, no. 1 (1999), 5.

105. Goschen to Grey, #264, 29 July 1914, in *British Documents on the Origins of the War: 1898–1914*, ed. G. P. Gooch and Harold Temperley (London: HMSO, 1926), 11: 171. Pourtalès to Auswärtiges Amt, #204, 31 July 1914, 7:10 pm, reel 1820 (Bd 11), T120, NARA. James Joll, "Politicians and the Freedom to Choose: The Case of July 1914," in *The Idea of Freedom: Essays in Honour of Isaiah Berlin,* ed. Alan Ryan (Oxford: Oxford University Press, 1979), 108–109.

106. Steiner, *Britain and the Origins of the War,* 226–227. William to His Majesty the King, Abschrift, 1 Aug. 1914, T120, reel 1820 (Bd 11), NARA. Aide mémoire (Lichnowsky account of conversation with Grey), 1 Aug. 1914 [847750–1], T120, reel 1820 (Bd 12), NARA. A. J. P. Taylor, interviewed in Norman F. Cantor, *Perspectives on the European Past: Conversations with Historians,* pt. 2 (New York: Macmillan, 1971), 274; John I. Knudson, *A History of the League of Nations* (Atlanta: Turner E. Smith, 1938), 137.

107. Joll, *Origins of the First World War,* 21.

108. The telegraph also provided a justification for pressuring others to make hasty decisions. When Nicholas Pasic, the Serbian prime minister, protested that the forty-eight-hour deadline made it impossible for him to call a full cabinet meeting in time, the Austro-Hungarian minister in Belgrade replied, "The return of the ministers in the age of railways, telegraph and telephone in a land of that size could only be a matter of a few hours." Albertini, *Origins of the War,* 2: 285; Tschirschky to Bethmann Hollweg, #251, 24 July 1914, T120, reel 1818 (Bd. 4), NARA.

109. Count Leopold von Berchtold said he had received information that the German government wanted Austria-Hungary to act with "the greatest speed" in its military operations "in order to prevent the interference of third parties if at all possible." Tschirschky to German foreign ministry, #105, 26 July 1914, T120, reel 1818 (Bd. 5), NARA. Kurt Riezler's diary, which appears to express Bethmann Hollweg's opinion, advocates a "quick fait accompli" followed by an effort to conciliate the Entente powers. *Kurt Riezler: Tagebücher, Augsätze, Dokumente,* ed. Karl Dietrich Erdmann (Göttingen: Vandenhoeck & Ruprecht, 1972), 11 July 1914, 185. The German ambassador in Vienna reported, "They have decided here to send out the declaration of war tomorrow, or the day after tomorrow at the latest, chiefly to frustrate any attempt at intervention." Tschirschky to German foreign ministry, 27 July 1914, in *Outbreak of the World War: German Documents Collected by Karl Kautsky,* ed. Karl Kautsky, Max Montgelas, and Walther Schücking (New York: Oxford University Press, 1924), no. 257, 243.

110. Albertini, *Origins of the War,* 2: 461; Jannen, *Lions of July,* 138.

111. Gregory, *Edge of Diplomacy,* 68–69. The following account also uses Albertini, *Origins of the War,* 3: 500–501; Bruce, *Silken Dalliance,* 183; Nicolson, *Sir Arthur Nicolson,* 425–426.

112. Gregory, *Edge of Diplomacy,* 70. Albertini, *Origins of the War,* 3: 500–502.

113. Jannen, *Lions of July,* 383; Lebow, *Between Peace and War,* 135–139; Young, *Prince Lichnowsky,* 106–112; Albertini, *Origins of the War,* 2: 503, 522–525.

114. "When negotiators have a competitive goal, time pressure enhances competitive behavior and lessens the likelihood of agreement." Peter J. Carnevale et al., "Time Pressure in Negotiation and Mediation," in *Time Pressure and Stress in Human Judgment and Decision Making,* ed. Ola Svenson and A. John Maule (New York: Plenum, 1993), 124. "The rapid plunge into actual war, amazingly fast by contemporary standards, made rational thought difficult." Steiner, *Britain and the Origins of the War,* 235.

6. The Zimmermann Telegram

1. Technically, it was not a "note" since Zimmermann did not intend for it to be presented in written form to the Mexican government. Goeppert Report, 4 Apr. 1917, 3, R16919, AA-PA.

2. Kurt Riezler sums up this military mind-set as expressed in Erich Ludendorff, its highest exponent: "Ludendorff, a brilliant energy perhaps a great strategist—but politically clueless and exceedingly ignorant, jittery, and precipitous. If left to his own devices, he would quickly plunge Germany into the abyss." Entry for 22 Nov. 1916, in *Kurt Riezler: Tagebücher, Aufsätze, Dokumente,* ed. Karl Dietrich Erdmann (Göttingen: Vandenhoeck & Ruprecht, 1972), 383.

3. Suzanne Tassier, *La Belgique et l'entrée en guerre des Etats-Unis, 1914–1917* (Brussels: La renaissance du livre, 1951); Count Bernstorff, *My Three Years in America* (London: Skeffington, [1920]), 7–8, 49, 108–109. Friedrich Katz, *The Secret War in Mexico: Europe, the United States and the Mexican Revolution* (Chicago: University of Chicago Press, 1981); Holger H. Herwig, *Germany's Vision of Empire in Venezuela, 1871–1914* (Princeton: Princeton University Press, 1986); Nancy Mitchell, *The Danger of Dreams: German and American Imperialism in Latin America* (Chapel Hill: University of North Carolina Press, 1999).

4. Count Bernstorff, *Memoirs of Count Bernstorff* (New York: Random House, 1936), 23, 27. Reinhard R. Doerries, *Imperial Challenge: Ambassador Count Bernstorff and German-American Relations, 1908–1917* (Chapel Hill: University of North Carolina Press, 1989), 11, 57–58; Bernstorff, *Three Years in America,* 349.

5. On the plagiarism, see *New York Times,* 25 Dec. 1914, 3; 26 Dec. 1914, 6; 28 Dec. 1914, 8. Bernhard Prince von Bülow, *Memoirs of Prince von Bülow,* trans. Geoffrey Dunlop (Boston: Little, Brown, 1931), 2: 575; entry for 28 Apr. 1912, in Baroness Spitzemberg, *Das Tagebuch der Baronin Spitzemberg,* ed. Rudolf Vierhaus (Göttingen: Vandenhoeck & Ruprecht, 1960), 543.

6. Martin Kitchen, *The Silent Dictatorship: The Politics of the German High Com-*

mand under Hindenburg and Ludendorff (New York: Holmes and Meier, 1976), 115; Martin Nassua, *"Gemeinsame Kriegführung. Gemeinsamer Friedensschluss." Das Zimmermann-Telegramm vom 13. Januar 1917 und der Eintritt der USA in den 1. Weltkrieg* (Frankfurt a. M.: Peter Lang, 1992), 38.

7. Telegrams of 5 Aug. 1914, 148–152, ADM 137/987, PRO; "Report on the standing Sub-Committee of the Committee of Imperial Defence: Submarine Cable Communications in Time of War," 11 Dec. 1911, CAB 38/19/56, PRO.

8. Memo from Gesellschaft für drahtlose Telegraphie m.b.H., 6 Oct. 1916, "Funken-telegraphische Verbindung mit Amerika," RM 5/1746, BA-MA. "Imperial Wireless Telegraphy Committee, 1919–1920" (London: HMSO, June 1920), 3–4, 800.72/60, 1910–1929, RG 59, NARA. Christopher Andrew, "Déchiffrement et diplomatie: Le cabinet noir du Quai d'Orsay sous la Troisième République," *Relations Internationales* 5 (1976), 59; memo of 15 Dec. 1916, 246, RM 5/1746, BA-MA.

9. Bernstorff, *Three Years in America,* 56. Haniel to Reichskanzler, 16 Aug. 1914, R901, Akt. 742, 96–97, BA. Robert Lansing, "Memorandum on Wireless Censorship by the Department of State," 16 Feb. 1916, 763.72/2414–1/2, M267, roll #25, NARA; Bernstorff to Bethmann Hollweg, 7 Mar. 1917, and testimony of Karl Gesemann, 23 Mar. 1917, R16919, AA-PA.

10. "Submarine to Take United States Mails," *New York Times,* 2 Nov. 1916, 4; "Mail by Submarine," *New York Times,* 3 Feb. 1917, 3. Patrick Beesly, *Room 40: British Naval Intelligence 1914–18* (London: Hamish Hamilton, 1982), 229. William F. Friedman and Charles J. Mendelsohn, *The Zimmermann Telegram of January 16, 1917 and Its Cryptographic Background* (1938; Laguna Hills, Calif.: Aegean Park, 1976), 6–7; Doerries, *Imperial Challenge,* 223; Bernstorff, *Three Years in America,* 223.

11. See, e.g., Bernstorff to Auswärtiges Amt, 23 Sept. 1914, R2437, AA-PA.

12. Lucius to Auswärtiges Amt, 8 Jan. 1915, R2438, AA-PA; Reichenau to Bethmann Hollweg, 20 Jan. 1915, R901, Aktenband 855, s. 185, BA; testimony of Zimmermann, 2d subcommittee, 7th sess., 6 Nov. 1919, in Carnegie Endowment for International Peace, *Official German Documents Relating to the World War* (Concord, N.H.: Rumford Press, 1923), 1: 479.

13. Foreign Office to Kohler, #88, 15 May 1915, ADM 116/1371, PRO. Friedman and Mendelsohn, *Zimmermann Telegram,* 10–11.

14. Friedman and Mendelsohn, *Zimmermann Telegram,* 13–14; Bernstorff, *Three Years in America,* 128–130, 140. Gerard to Bryan, 18 Dec. 1915, and State Department to Gerard, 20 Dec. 1915, 763.72/2311; Lansing to Bernstorff, Washington, 20 Dec. 1915, 763.72/2328–1/2; and Lansing to Gerard, 21 Dec. 1915, 763.72/2324, all at M367, reel 25, NARA.

15. Robert Lansing, *War Memoirs of Robert Lansing* (Indianapolis: Bobbs-Merrill, 1935), 227.

16. Bernstorff, *Memoirs,* 142, 143. Bethmann Hollweg repeatedly complained to the U.S. ambassador to Germany about his difficulties in communicating with Bernstorff. See [Gerard] to Lansing, 11 May 1916, 163.72/2687, M367, reel 28,

NARA. Doerries, *Imperial Challenge,* 107. For an analysis of the *Lusitania* crisis at variance with mine, see Arthur S. Link, *Wilson: The Struggle for Neutrality, 1914–1915* (Princeton: Princeton University Press, 1960), 375–377.

17. Bernstorff to German foreign ministry, 27 Jan. 1917, quoted in Doerries, *Imperial Challenge,* 216–217.

18. Lansing, *War Memoirs,* 210–212.

19. Johannes Hürter, "Die Staatssekretäre des Auswärtigen Amts im Ersten Weltkrieg," in *Der Erste Weltkrieg: Wirkung, Wahrnehmung, Analyse,* ed. Wolfgang Michalka (Munich: Piper, 1994), 236; Jagow to Bernstorff, 2 Sept. 1919, in Bernstorff, *Memoirs,* 164–165.

20. Gerhard Ritter, *The Sword and the Scepter: The Problem of Militarism in Germany* (Coral Gables, Fla.: University of Miami Press, 1972), 3: 290, 320; Kitchen, *Silent Dictatorship,* 120; James W. Gerard, *My Four Years in Germany* (New York: Doran, 1917), 376–377. Hürter, "Die Staatssekretäre," 250n46.

21. Ritter, *Sword and Scepter,* 3: 213; Hürter, "Die Staatssekretäre," 224–225. Zimmermann had been considered for the post of foreign secretary in December 1912, but his bourgeois background, among other reasons, made him unsuitable for the post at that time. Entry of 9 Jan. 1913, in Baroness Spitzemberg, *Tagebuch,* 554; Joseph C. Grew, *Turbulent Era: A Diplomatic Record of Forty Years, 1904–1945,* ed. Walter Johnson (Boston: Houghton Mifflin, 1952), 1: 220.

22. Hürter, "Die Staatssekretäre," 236, 224; Gordon Craig, *Germany: 1866–1945* (New York: Oxford University Press, 1978), 378. Nassua, *"Gemeinsame Kriegführung."* Zimmermann eventually proved too much a diplomat for the taste of the army leaders, who forced him to resign on 7 August 1917.

23. Arthur S. Link, *Wilson: Campaigns for Progressivism and Peace, 1916–1917* (Princeton: Princeton University Press, 1965), 433–436, 344; Katz, *Secret War in Mexico,* 353.

24. Link, *Wilson: Campaigns for Progressivism and Peace,* 433–434, 344; Katz, *Secret War in Mexico,* 541–542, 350, 345, 348–350. See also statements of Eduard David and Zimmermann, 5 Mar. 1917, in Hans Peter Hanssen, *Diary of a Dying Empire,* trans. Oscar Osburn Winther, ed. Ralph H. Lutz et al. (Bloomington: Indiana University Press, 1955), 176, 178.

25. I have found no evidence that the telegram traveled via the Swedish route, although many historians have claimed that it did. Given existing German practices regarding important diplomatic cables, this claim may be warranted even though there is no clear evidence to support it. Friedman and Mendelsohn, *Zimmermann Telegram,* 11; David Kahn, *The Codebreakers: The Story of Secret Writing* (New York: Scribner's, 1996), 284–285, 1039n284. Nonetheless, I believe the claim probably testifies to the effectiveness of British misinformation (intended to conceal British espionage against U.S. diplomacy). Sweden could not have directly conveyed the message to the German legation in Mexico City unless the message had been in code 13040, whereas the evidence indicates that the telegram reached the Americas in code 0075. The Goeppert investigation suggests that Sweden did not transmit it

to the Washington embassy either, since that post received only one copy of the telegram, which the U.S. State Department transported. (Zimmermann sent a second telegram, modifying the first, on 5 February 1917. It traveled via Swedish cable. See Goeppert Report, 2.) The slowness of the Swedish route—sending a telegram and receiving a reply could take as long as a week—might explain the German reluctance to use it to convey a time-sensitive message when the faster State Department route was available. Many historians have also repeated the claim of Burton J. Hendrick, in *The Life and Letters of Walter Hines Page* (Garden City, N.Y.: Doubleday, Page, 1925), 3: 335–342, and Barbara Tuchman, in *The Zimmermann Telegram* (New York: Macmillan, 1966), 146, that the message went out on wireless. This is almost certainly incorrect. The U.S. government required Germany to submit the cleartext of all encrypted transatlantic radiograms sent to the United States, as well as all codebooks used to encrypt them. The U.S. government would have censored the telegram had it been sent by radio. It is equally unlikely that the German government sent Zimmermann's message directly to Mexico via radio. Although, in November 1916, the Mexican government advocated the establishment of radio communications between Germany and Mexico, the two governments were still in the process of constructing this link in 1917. See Nassua, *"Gemeinsame Kriegführung,"* 23; Friedman and Mendelsohn, *Zimmermann Telegram,* 14; Goeppert Report, 13.

26. Even when the State Department code used numbers, it employed them in groups of five, whereas Germany's code 0075 used numbers in groups of four. Other German codes, such as 13040, used a mixture of three-, four-, and five-digit groups, which were easily distinguishable from the American code. Likewise, British cryptanalysts became suspicious when they noticed code groups in Swedish telegrams that resembled German diplomatic codes. Kahn, *Codebreakers,* 286, 284; Friedman and Mendelsohn, *Zimmermann Telegram,* 12–13.

27. The difficulty caused by garbling indicates that the German embassy did not receive a second, Swedish-conveyed copy of the telegram. Testimony of Prince von Hatzfeld, 2, 19 Mar. 1917, R16919, AA-PA. Goeppert Report, 6. *New York Times,* 2 Nov. 1916, 1.

28. Many years later Nigel de Grey, who helped decipher the Zimmermann telegram, said that he first received it in code 13040. But British intelligence would have initially encountered the Zimmermann telegram in 0075 when the State Department sent it from Berlin to Washington. Perhaps de Grey's memory is faulty and he first encountered the Zimmermann telegram in 0075, made some headway against it (as I argue), and filled in the gaps after British intelligence acquired a copy of it in 13040 (probably in Mexico City). If de Grey's memory is accurate, this implies that British intelligence regularly acquired telegrams sent to the German legation in Mexico City, obtained a copy of the Zimmermann telegram (without realizing its importance) and decoded it in 13040, and in fact made little headway against 0075 until later. See Nigel de Grey, "Zimmermann Telegram: A Footnote to Friedman's Account," 31 Oct. 1945, in David Kahn, "Edward Bell and His Zimmermann Tele-

gram Memoranda," *Intelligence and National Security* 14, no. 3 (Autumn 1999), 153–156; Friedman and Mendelsohn, *Zimmermann Telegram,* 20–21.

29. Kahn, *Codebreakers,* 286. Every complete sentence of the Zimmermann telegram ended with a verb. Goeppert Report, 2–4.

30. Kahn, *Codebreakers,* 283, 276; Beesly, *Room 40,* 204, 36–38; Christopher Andrew, *Her Majesty's Secret Service: The Making of the British Intelligence Community* (New York: Elisabeth Sifton, 1986), 91.

31. Kahn, "Edward Bell," 155. Robert Lansing memo, 4 Mar. 1917, in *Papers of Woodrow Wilson,* 41: 325.

32. Admiral Sir William James, *The Eyes of the Navy: A Biographical Study of Admiral Sir Reginald Hall* (London: Methuen, 1956), 134–136, 140–141; Katz, *Secret War in Mexico,* 358. "Admiral Hall on the Zimmermann Telegram," *World's Work* 51, no. 6 (Apr. 1926), 579.

33. Page to Lansing, #5747, London, 24 Feb. 1917, 862.20212/69, M336, reel 55, NARA.

34. Page to Lansing, #5747, London, 24 Feb. 1917, 862.20212/69, M336, roll #55, NARA. Lansing memo, 4 Mar. 1917, 322. In accordance with federal law protecting the confidentiality of telegrams, Western Union initially refused the U.S. government's request for the Zimmermann dispatch, before bowing to intense lobbying. Ibid., 323; Tuchman, *Zimmermann Telegram,* 171–174.

35. Friedman and Mendelsohn, *Zimmermann Telegram,* 28–29; "Admiral Hall on the Zimmermann Telegram," 579. Christopher Andrew, *For the President's Eyes Only: Secret Intelligence and the American Presidency from Washington to Bush* (New York: HarperCollins, 1995), 45.

36. Stellvertretender Generalstab der Armee, 13 Mar. 1917, R16919, AA-PA. Andrew, *Her Majesty's Secret Service,* 114. Otto von Kellenbach to Reichskanzler Michaelis, 7 Sept. 1917, and German consulate in Basel, 25 Sept. 1917, R16919, AA-PA. *The Intimate Papers of Colonel House,* ed. Charles Seymour (Boston: Houghton Mifflin, 1926), 452–453; Friedman and Mendelsohn, *Zimmermann Telegram,* 4n11. Lansing memo, 4 Mar. 1917, 325.

37. Page to Lansing, #5747, 24 Feb. 1917, 862.20212/69, M336, roll #55, NARA. Ross Gregory, *Walter Hines Page: Ambassador to the Court of St. James's* (Lexington: University Press of Kentucky, 1970), 192.

38. Lansing, *War Memoirs,* 227; Link, *Wilson: Campaigns for Progressivism and Peace,* 345; Polk diary, 25 Feb. 1917, folder 0002, box 17, ser. II, #656, Frank L. Polk Papers, Yale University Manuscripts and Archives.

39. Lansing memo, 4 Mar. 1917, 323, 324. Draft instruction to Page, undated and unsigned, folder 322, box 26, ser. III, #656, Polk Papers. In fact, Nigel de Grey did most of the deciphering while the American official looked on. Because de Grey had brought the wrong code book, he pretended to decipher the telegram while actually repeating it from memory. Kahn, *Codebreakers,* 294; Beesly, *Room 40,* 223; Andrew, *Her Majesty's Secret Service,* 113.

40. Zimmermann appears to have been motivated by two beliefs: that the United States probably possessed convincing proof of the offer, making a denial useless; and that publicity surrounding the offer to Mexico might deter the United States from entering the war. See Katz, *Secret War in Mexico,* 372.

41. Address to joint session of Congress, 2 Apr. 1917, *Papers of Woodrow Wilson,* 41: 525. On the U.S. entrance into the war, see, e.g., Ernest R. May, *The World War and American Isolation, 1914–1917* (Cambridge, Mass.: Harvard University Press, 1959); Ragnhild Fiebig-von Hase, "Der Anfang vom Ende des Krieges: Deutschland, die USA und die Hintergründe des amerikanischen Kriegseintritts am 6. Apr. 1917," in *Der Erste Weltkrieg,* ed. Michalka, 125–158.

42. House to Wilson, 27 Feb. 1917, and Wilson to House, 26 Feb. 1917, in *Papers of Woodrow Wilson,* 41: 296–297, 288. On 28 Jan. 1917 Lansing predicted in a memo: "Sooner or later the die will be cast and we will be at war with Germany . . . When that time comes, . . . it will come because of German folly . . . I hope that those blundering Germans will blunder soon." Lansing, *War Memoirs,* 208.

43. John Milton Cooper, *The Warrior and the Priest: Woodrow Wilson and Theodore Roosevelt* (Cambridge, Mass.: Harvard University Press, 1983), 241–242, 318. On Wilson's sensitivity to affronts, see Bernstorff, *Three Years in America,* 53.

44. Link, *Wilson: Campaigns for Progressivism and Peace,* 346. Hull and Addams in *Papers of Woodrow Wilson,* 41: 305n3.

45. "Washington Officials Stunned," *New York Times,* 1 Feb. 1917, 2. Lansing memo, 4 Mar. 1917, 327. Lansing, *War Memoirs,* 225.

46. Link, *Wilson: Campaigns for Progressivism and Peace,* 357, 359; Samuel R. Spencer Jr., *Decision for War, 1917: The Laconia Sinking and the Zimmermann Telegram as Key Factors in the Public Reaction against Germany* (Rindge, N.H.: Richard R. Smith, 1953), 79; Nassua, *"Gemeinsame Kriegführung,"* 114.

47. Entry for 4 Mar. 1917, Riezler, *Tagebücher,* 412. Statement of Eduard David, 5 Mar. 1917, in Hanssen, *Diary of a Dying Empire,* 177. Katz, *Secret War in Mexico,* 367–373, 378. Lansing memo, 4 Mar. 1917, 327; statement of Zimmermann, 5 Mar. 1917, in Hanssen, *Diary of a Dying Empire,* 179. On newspaper opinion, see comments of the Prussian legation in Karlsruhe, 5 Mar. 1917, R16919, AA-PA; and Langhorne to Lansing, #743, 5 Mar. 1917, 763.72/3415; Langhorne to Lansing, #746, 6 Mar. 1917, 763.72/3428; Morris to Lansing, #210, 7 Mar. 1917, 763.72/3449; Egan to Lansing, #488, 7 Mar. 1917, 763.72/3452; and Stovall to Lansing, #625, 8 Mar. 1917, 763.72/3456, all at M367, reel #32, NARA.

48. Statement of Zimmermann, 22 Feb. 1917, in Hanssen, *Diary of a Dying Empire,* 175; Doerries, *Imperial Challenge,* 229. *New York World* to Bernstorff, 2 Mar. 1917, R16919, AA-PA. Michahelles to Auswärtiges Amt, #171, 10 Mar. 1917, and Rantzau to Auswärtiges Amt, #416, 12 Mar. 1917, R16919, AA-PA.

49. Stovall to Lansing, #603, 3 Mar. 1917, 763.72/3418, M367, #32, NARA.

50. Statement of Zimmermann, 5 Mar. 1917, in Doerries, *Imperial Challenge,* 230, and in Hanssen, *Diary of a Dying Empire,* 177. Testimony of Bernstorff, 2d sub-

committee, 3d sess., 23 Oct. 1919, in Carnegie Endowment, *Official German Documents,* 1: 316; Bernstorff, *Three Years in America,* 345–353. Morgen Ausgabe, "Graf Bernstorff," *Kreuz-Zeitung,* 17 Feb. 1921, 2.

51. Zimmermann memo, 17 Mar. 1917, R16919, AA-PA. Goeppert Report, 6–7, 13, 4, 16.

52. Goeppert Report, 14–15. Stellvertretender Generalstab der Armee to Auswärtiges Amt, 27 May 1917, R16919, AA-PA. Bernstorff to Auswärtiges Amt, #171, 10 Mar. 1917, and #416, 12 Mar. 1917, R16919, AA-PA; Katz, *Secret War in Mexico,* 375, 377.

53. Scott D. Sagan, *The Limits of Safety: Organizations, Accidents, and Nuclear Weapons* (Princeton: Princeton University Press, 1993), 208.

54. Doerries, *Imperial Challenge,* 230, 237n36, 365n190. Testimony of Bernstorff, 2d subcommittee, 3d sess., 23 Oct. 1919, in Carnegie Endowment, *Official German Documents,* 1: 310–311.

55. Bernstorff, *Memoirs,* 27, 49, 285–288. After the murder of Germany's first Jewish foreign minister, Walter Rathenau, in 1922, Bernstorff wrote to the *Frankfurter Zeitung:* "In England it never occurred to anyone to attack the great statesman Disraeli on the ground of his race. Our Republic needs men of firm character and resolution, who will set themselves against this antisemitic fever before it leads to further disaster." Bernstorff, *Memoirs,* 287.

56. Nassua, *"Gemeinsame Kriegführung,"* 20; Katz, *Secret War in Mexico,* 354. Chiffrierbureau memo, 24 Feb. 1917, R16919, AA-PA. Bernstorff, *Three Years in America,* 325.

57. Charles J. Mendelsohn, *Studies in German Diplomatic Codes Employed during the World War* (Washington: GPO, 1937), 69–71, RG 457, box 745, NARA. Friedman and Mendelsohn, *Zimmermann Telegram,* 26. Affidavit of W. Reginald Hall, in Friedman and Mendelsohn, *Zimmermann Telegram,* 30. Goeppert Report, 11.

58. Mendelsohn, *Studies in German Diplomatic Codes.* Goeppert Report, 11. Affidavit of W. Reginald Hall, 30. At the German embassy in Washington, "the cipher was not changed as often as would have been the case under normal conditions. In all probability, if communications had not been interrupted, we would have received new ciphers every month or every other month, so that they could not have been found out so easily." Testimony of Bernstorff, 2d subcommittee, 7th sess., 6 Nov. 1919, in Carnegie Endowment, *Official German Documents,* 1: 480.

59. Beesly, *Room 40,* 28–29.

60. The experience of the Zimmermann telegram did finally convince the Auswärtiges Amt to retire code 13040. Katz, *Secret War in Mexico,* 614n115.

61. The codes to which he referred (he named 0053, 0075, and 0097) were the "Lotteriechiffres," codes in which the numbers were not in any sort of alphabetical order. Romberg to Reichskanzler, 18 Feb. 1918, R2589, AA-PA.

62. Simon Singh, *The Code Book: The Evolution of Secrecy from Mary Queen of Scots to Quantum Cryptography* (New York: Doubleday, 1999), 107.

63. Goeppert Report, 8. "Washington Refutes a Charge by Germans," *New York Times,* 20 Sept. 1917, 3.

64. Andrew, *Her Majesty's Secret Service,* 97–108. Hall later declared: "Owing to the paramount importance of our having for the use of the British Navy the information contained in the messages regarding the movement of German ships it was imperative that we should avoid if possible, disclosing to the Germans the fact that we were reading their communications to this extent. Hence it was impossible for us at the time to make full use of all the information which was before us." Affidavit of W. Reginald Hall, 31. James, *Eyes of the Navy,* 138; Beesly, *Room 40,* 17–18, 20.

65. Baron J. de Szilassy, *Traité pratique de diplomatie moderne* (Paris: Payot, 1928), 164, citing François de Callières, an eighteenth-century commentator on diplomacy. One could see telegraphic signals intelligence as a precursor to later forms of intelligence that undermined secrecy, reduced the danger of a surprise attack, and thereby promoted stability during the Cold War. See John Lewis Gaddis, "Learning to Live with Transparency: The Emergence of a Reconnaissance Satellite Regime," in Gaddis, *The Long Peace: Inquiries into the History of the Cold War* (Oxford: Oxford University Press, 1987), 195–214.

66. "Washington Refutes a Charge by Germans," *New York Times,* 20 Sept. 1917, 3; Goeppert Report, 8. Nelson Page to Lansing, 3 Mar. 1917, 763.72/3414; Guthrie to Lansing, 2 Mar. 1917, 763.72/3425; and Guthrie to Lansing, 6 Mar. 1917, 763.72/3426, all at M367, reel 32, NARA.

67. Viereck quoted in Katz, *Secret War in Mexico,* 360.

68. A. J. P. Taylor, "Bismarck's Morality," in Taylor, *Europe: Grandeur and Decline* (New York: Penguin, 1979), 94–95; Sebastian Haffner, "Otto von Bismarck," in *Preussische Profile,* ed. Sebastian Haffner and Wolfgang Venohr (Frankfurt: Ullstein, 1986), 158, 163.

69. Bernstorff, *Memoirs,* 127.

70. Testimony of Prince von Hatzfeld, 2–3, and testimony of Heinrich Schaffhausen, 2, both 19 Mar. 1917, R16919, AA-PA; Goeppert Report, 10; Friedman and Mendelsohn, *Zimmermann Telegram,* 16. Henry F. Schorreck, "The Telegram That Changed History," *Cryptologic Spectrum* (Summer 1970), 23, RG 457, SRH 234, entry 9002, NARA; John J. McCusker, "Comparing the Purchasing Power of Money in the United states (or Colonies) from 1665 to Any Other Year Including the Present." Economic History Services, 2001, *www.eh.net/hmit/ppowerusd/.*

7. Technical and Economic Factors

1. Tocqueville, "Extrait d'une circulaire in date du 27 Septembre 1849" in Circulaire #76, 1 Nov. 1850, Comptabilité, Cartons, 9, MAE, Paris. Despite colorful stories about stolen diplomatic correspondence, such espionage had relatively little effect upon foreign policy before the telegraph. Sympathizers or spies in opposing governments provided a simpler and more effective means of gathering intelligence than did efforts at communications intelligence. Signals intelligence came into its own

when humans entered first the electric and then the electronic age. See Charles Carter, *The Western European Powers, 1500–1700* (Ithaca: Cornell University Press, 1971), 236–239.

2. Ralph E. Weber, *Masked Dispatches: Cryptograms and Cryptology in American History, 1775–1900* (Fort Meade: National Security Agency, Center for Cryptologic History, 1993), 207. Even the United States nationalized its system during the First World War and, after the war, made arrangements with Western Union and the Postal Telegraph Company to obtain messages sent by foreign envoys. David Kahn, "The Annotated *The American Black Chamber,*" *Cryptologia* 9 (Jan. 1985), 20.

3. Confidential circular, T. F. Bayard, 13 Oct. 1886, vol. 3, E726 (Circulars of the Department of State), RG 59, NARA. "Ciphers and Cipher Keys," *Living Age* 333 (15 Sept. 1927), 493. David Alvarez, "Faded Lustre: Vatican Cryptography, 1815–1912," *Cryptologia* 20, no. 2 (Apr. 1996), 119–126; David Alvarez, "Vatican Communications Security, 1914–1918," *Intelligence and National Security* 7, no. 4 (1992), 447.

4. Although the words "code" and "cipher" are sometimes used interchangeably, a code is generally a symbol (such as a word or number) used to represent a phrase, a syllable, or, most often, a *word.* A cipher uses the *letter,* rather than the word, as the basic unit. A cipher thus tends to be more systematic. See David E. Newton, *Encyclopedia of Cryptology* (Santa Barbara: ABC-Clio, 1997). Nineteenth-century foreign ministries often used the term "cipher" to designate what we would call a code.

5. Christopher Andrew, "Déchiffrement et Diplomatie: Le cabinet noir du Quai d'Orsay sous la Troisième République," *Relations Internationales* 5 (1976), 43–44. David Schimmelpenninck van der Oye, "Tsarist Codebreaking: Some Background and Some Examples," *Cryptologia* 22, no. 4 (Oct. 1998), 342–353; David Kahn, *The Codebreakers: The Story of Secret Writing* (New York: Scribner, 1996), 621; Christopher Andrew and Oleg Gordievsky, *KGB: The Inside Story* (New York: Harper Perennial, 1991), 29–30. David Schimmelpenninck van der Oye, "A First Look at Russia's Codebreakers: A Book Review," *Cryptologia* 21, no. 1 (Jan. 1997), 39.

6. Reichenau telegram, 29 Oct. 1914, R11104, AA-PA; James Joll, *The Origins of the First World War* (London: Longman, 1989), 12.

7. Extract from *Truth,* 22 Dec. 1898, 1567, R11104, AA-PA.

8. George von Lengerke Meyer to Theodore Roosevelt, 5 July 1905, *Theodore Roosevelt Papers* (microfilm, Harvard University), ser. I, reel 56; Ralph E. Weber, *United States Diplomatic Codes and Ciphers, 1775–1938* (Chicago: Precedent, 1979), 246. Willisch, 28 Nov. 1904, R11104, AA-PA. Christopher Andrew and Keith Neilson, "Tsarist Codebreakers and British Codes," in *Codebreaking and Signals Intelligence,* ed. Christopher Andrew (London: Frank Cass, 1986), 8.

9. H. J. Bruce, *Silken Dalliance* (London: Constable, 1947), 92–93, 82–83. Andrew and Neilson, "Tsarist Codebreakers," 10.

10. Extract from *Truth,* 22 Dec. 1898, 1567, R11104, AA-PA. Fürst von Radolin to von Bülow, 20 Dec. 1898, R11104, AA-PA. Andrew and Neilson, "Tsarist Codebreakers," 6.

11. Andrew, "Déchiffrement et Diplomatie," 42, 43, 50; Kahn, *Codebreakers,* 255;

Weber, *Masked Dispatches,* 223. Pierre Renouvin, *Histoire des relations internationales* (Paris: Hachette, 1955), 6: 250.

12. Octave Homberg, *Les coulisses de l'histoire: Souvenirs, 1898–1928* (Paris: Librairie Arthème Fayard, 1938), 43–45; Christopher Andrew, *Théophile Delcassé and the Making of the Entente Cordiale: A Reappraisal of French Foreign Policy, 1898–1905* (London: Macmillan, 1968), 290, 297, 300–301.

13. Undated and unsigned memo, probably first half of 1918, AS 1309.11, zu AS 1291, R2590, AA-PA. "Aufzeichnung über Herstellung und Versendung von Chiffres seit dem Jahre 1911," 12 Mar. 1917, R16919, AA-PA. Andrew, "Déchiffrement et Diplomatie," 53–55. Freiherr Wilhelm von Schoen, *The Memoirs of an Ambassador: A Contribution to the Political History of Modern Times,* trans. Constance Vesey (London: George Allen and Unwin, 1922), 196; Pierre Renouvin, *The Immediate Origins of the War* (New Haven: Yale University Press, 1928), 269; Luigi Albertini, *The Origins of the War of 1914,* trans. and ed. Isabella M. Massey (London: Oxford University Press, 1953), 3: 214. Andrew and Gordievsky, *KGB,* 30.

14. Andrew, "Déchiffrement et Diplomatie," 46–50, 55; Andrew, *Théophile Delcassé,* 72–73; Andrew, "Codebreakers and Foreign Offices," 34.

15. Andrew, *Théophile Delcassé,* 98–99. Kahn, *Codebreakers,* 254–262; Andrew, "Déchiffrement et Diplomatie," 48. Homberg, *Les coulisses de l'histoire,* 39–40.

16. Kenneth Bourne, "The Foreign Office under Palmerston," in *The Foreign Office, 1782–1982,* ed. Roger Bullen (Frederick, Md.: University Publications of America, 1984), 35. Harald Hubatschke, "Die amtliche Organisation der geheimen Briefüberwachung und des diplomatischen Chiffrendienstes in Österreich (von den Anfängen bis etwa 1870)," *Mitteilungen des Instituts für Österreichische Geschichtsforschung* 83 (1975), 352–413; Kahn, *Codebreakers,* 163–165; Alvarez, "Faded Lustre," 106, 411. Nonetheless, the Russians found the Austrians a worthy adversary. See Schimmelpenninck, "Tsarist Codebreaking," 350.

17. "Ciphers and Cipher Keys," *Living Age,* 494; Kahn, *Codebreakers,* 264.

18. Kahn, *Codebreakers,* 263; Andrew and Gordievsky, *KGB,* 29; Michael B. Miller, *Shanghai on the Métro: Spies, Intrigue, and the French between the Wars* (Berkeley: University of California Press, 1994), 22–23.

19. Willisch, memo, 26 Aug. 1901, R11104, AA-PA. Robert K. Massie, *Dreadnought: Britain, Germany, and the Coming of the Great War* (New York: Random House, 1991), 726.

20. Propp, memo, 13 Jan. 1916, R2589, AA-PA. Charles J. Mendelsohn, *Studies in German Diplomatic Codes Employed during the World War* (Washington: GPO, 1937), 71, RG 457, box 745, NARA.

21. Dennis Mack Smith, *Mazzini* (New Haven: Yale University Press, 1994), 41–43. Christopher Andrew, *Her Majesty's Secret Service: The Making of the British Intelligence Community* (New York: Viking, 1986), 85. Weber, *U.S. Diplomatic Codes and Ciphers,* 227. John Ferris, "Before 'Room 40': The British Empire and Signals Intelligence, 1898–1914," *Journal of Strategic Studies* 12, no. 4 (Dec. 1984), 431–457. Neil Hart, *The Foreign Secretary* (Lavenham, Suffolk: Terence Dalton, 1987), 149.

22. John Tilley and Stephen Gaselee, *The Foreign Office* (London: Putnam's, 1933), 148. Lord Granville to Mr. Gladstone, 23 Oct. 1880, in *The Political Correspondence of Mr. Gladstone and Lord Granville, 1876–1886,* ed. Agatha Ramm (Oxford: Clarendon, 1962), 1: 207.

23. Italy also lagged in cryptography before the war. David Alvarez, "Italian Diplomatic Cryptanalysis in World War I," in *Selections from Cryptologia: History, People, and Technology,* ed. Cipher A. Deavours et al. (Boston: Artech, 1998), 181–190.

24. The lack of a cipher in foreign posts hindered the State Department's communications with U.S. diplomats in Britain before the War of 1812 and in Brazil before American entry into the First World War. See Russell to Monroe, 1 May 1812, M30, reel 14, NARA; Monroe to Russell, 5 May 1812, M77, reel 2, NARA; Belden to Lansing, #863, 22 Jan. 1917, M367, reel 32, NARA.

25. Hamilton Fish to Robert C. Schenck, 16 June 1872, Hamilton Fish Papers, Container 348, LC. Hoffman Philip to William Phillips, 18 Jan. 1915, 119.2/124 1/2, 1910–29, RG 59, NARA. See also F. M. Gunther to W. Phillips, 11 June 1915, 119.2/147; and U.S. embassy in London to secretary of state, 15 Feb. 1916, M367, reel 25 (763.72/2408), NARA. Kahn, *Codebreakers,* 192–195. Weber, *Masked Dispatches,* 8.

26. Bigelow to Seward, #352, 3 Aug. 1866, and #375, 12 Oct. 1866, M34, reel 64, NARA. See also Charles Francis Adams to Seward, #1271, 21 Sept. 1866, M30, reel 88, NARA.

27. Seward to Bigelow, #512, 21 Aug. 1866, M77, reel 58, NARA. All of this occurred before the abolition of the British black chamber in 1844. Weber, *U.S. Diplomatic Codes and Ciphers,* 209–210.

28. Weber, *Masked Dispatches,* ch. 17. *New York Herald,* 7 Dec. 1866, 8. Weber, *U.S. Diplomatic Codes and Ciphers, 1775–1938,* 214. The State Department later sought to prohibit the practice of sending messages both in code and in clear. See confidential circulars, T. F. Bayard, 13 Oct. 1886, vol. 3; Elihu Root, 23 Mar. 1907, vol. 6; and Alvey A. Adee, 14 Oct. 1914, vol. 7, all at E726, RG 59, NARA.

29. Weber, *Masked Dispatches,* 216–217, 223. As a temporary expedient, the State Department supplemented the Red Code with the Slater Code, a public code, during the war with Spain. See Thomas W. Cridler, 8 Jan. 1900, vol. 5, E726, RG 59, NARA; Buck to Carr, 5 Aug. 1912, 119.25/82, 1910–1929, RG 59, NARA.

30. George von Lengerke Meyer to Theodore Roosevelt, 5 July 1905, in *Theodore Roosevelt Papers,* ser. I, reel 56 (microfilm, Harvard University). Perhaps in response to the theft of the Blue Code in Russia, the State Department reemphasized the prohibition against entrusting code books to foreigners. See Elihu Root, "Confidential," 16 Jan. 1909; and Robert Bacon, 27 Nov. 1908, vol. 7, E726, RG 59, NARA.

31. George Barclay Rives to Elihu Root, 26 Sept. 1907, file no. 4213/34, M862, reel 381, NARA. Most material on the Bucharest theft is in file no. 4213, M862, reel 381, NARA; see also Nelson O'Shaughnessy to John R. Buck, 8 Oct. 1913, 119.25/121, and O'Shaughnessy to Bryan, 12 Apr. 1914, 119.23/1, RG 59, 1910–29, NARA. Hill to Root, 21 Dec. 1908, file no. 16682, M862, reel 971, NARA.

32. Weber, *Masked Dispatches,* 231–232. 119.25/81–124, 1910–29, RG 59, NARA;

Frederick C. Penfield to secretary of state, 19 Feb. 1916, #1328, 119.252/185, 1910–29, RG 59, NARA.

33. Francis to Lansing, 24 May 1917, 119.252/196; minute by Salmon, 4 Dec. 1917, 119.252/217; A. Longfellow Milmore, 11 Dec. 1917, 119.252/222; and Morris to Lansing, 19 July 1918, 119.252/256, all at 1910–29, RG 59, NARA.

34. Dean, Buck, and Martin to Alvey Adee, 23 Mar. 1908, file no. 11311, M862, reel 767, NARA. John E. Osborne to Penfield, #825, 22 Mar. 1916, 119.252/185, 1910–29, RG 59, NARA.

35. Francis to Lansing, 24 May 1917, 119.252/196; Wm Whiting Andrews to Lansing, 21 May 1917, #77, 119.252/199; D.A.S. to Phillips, 19 Dec. 1917, 119.252/220; Grant Smith to Lansing, 119.252/266; and Grant Smith, #1708, 3 Sept. 1918, 119.252/270, all at 1910–29, RG 59, NARA. I have found no evidence that the U.S. legation in Romania lost its copy of the Green Code before the First World War. On this point, historians have been misled by Allen Dulles's partially accurate account of the loss of the Blue Code at the U.S. legation in Bucharest, a story that he had heard as a young foreign service officer. See Weber, *U.S. Diplomatic Codes and Ciphers,* 1, 17–18n1–2; Allen Dulles, *The Craft of Intelligence* (New York: Harper and Row, 1963), 73–74.

36. James Thurber, "Exhibit X," *New Yorker,* 6 Mar. 1948, 26. Kahn, *Codebreakers,* 490. Thurber, when working as a code clerk at the Paris embassy, once took a code book home to study after hours. His ability to do so implies a laxity on the part of the State Department in preserving its communications security. See Thurber, "Exhibit X," 26; Harrison Kinney, *James Thurber: His Life and Times* (New York: Henry Holt, 1995), 194.

37. See 8 June 1914, ADM 12/1530; 13 Dec. 1915, ADM 12/1546B; telegrams of 15 Jan., 17 Jan., and 27 Apr. 1916, ADM 12/1567A; 13 Jan. 1915, ADM 12/1546B; and 31 Dec. 1915, ADM 12/1546B, all at index heading 93, PRO.

38. Chief of naval operations to All Ships and Stations concerning Compromise of Naval Code F-3, 9 Mar. 1925, in "From the Archives: Compromise of Navy Code," *Cryptologia* 13, no. 4 (Oct. 1989), 381.

39. Andrew, "Déchiffrement et Diplomatie," 56. Admiral Sir William James, *The Eyes of the Navy: A Biographical Study of Admiral Sir Reginald Hall* (London: Methuen, 1956), 156–158; Beesly, *Room 40,* 237–241; Harold F. Peterson, *Argentina and the United States, 1810–1960* (New York: State University of New York Press, 1964), 311–315.

40. Tilley and Gaselee, *Foreign Office,* 148–149. Speaking of a slightly later period, Jacques Baeyens, a French diplomat, maintained that in general, the decryptions that he saw "concerned trifles" and were "without interest." He also quotes approvingly a code clerk who complained that the coded telegrams over which he toiled were too long, without interest, and would anyway be appearing tomorrow in the newspapers. Baeyens, *Au Bout du Quai: Souvenirs d'un retraité des postes* (Paris: Fayard, 1975), 224, 217. For the complaints of a U.S. diplomat charged with laboring over "cipher telegrams of very little importance," see diary of John MacMurray,

3 and 24 Feb. 1909, John van Antwerp MacMurray Collection, box 15, Princeton University.

41. On disadvantages of secrecy, see Stephen Van Evera, *Causes of War: Power and the Roots of Conflict* (Ithaca: Cornell University Press, 1999), 141–142; Daniel Patrick Moynihan, *Secrecy: The American Experience* (New Haven: Yale University Press, 1998); Monteagle Stearns, *Talking to Strangers: Improving American Diplomacy at Home and Abroad* (Princeton: Princeton University Press, 1996), 115, 126; Ithiel de Sola Pool's commentary in *Computers, Communications and the Public Interest,* ed. Martin Greenberger (Baltimore: Johns Hopkins University Press, 1971), 95.

42. Even in France under the Third Republic, domestic surveillance, driven by a fear of internal subversion, provided the greatest rationale for the activities of the Cabinet Noir. Douglas Porch, *The French Secret Services: From the Dreyfus Affair to the Gulf War* (New York: Farrar, Straus and Giroux, 1995), 20–21, 25.

43. Jim Reeds, "Data Compression—For Telegraphy," *Antenna: Newsletter of the Mercurians, in the Society for the History of Technology* 9 (Apr. 1997), 10; Kahn, *Codebreakers,* 189–190, 837–853; Kraetke (of Reichspostamt) to German foreign secretary, 25 Dec. 1908, R901, Akt. 710, 56, BA. Procter and Gamble found it uneconomical to sell cottonseed oil in Turkey unless it could transmit its offers in code, which the Ottoman government was loath to allow due to security fears. Procter and Gamble Co. to secretary of state, 4 Jan. 1908, File 11311 (1906–1910), M862, reel 132, NARA.

44. Weber, *Masked Dispatches,* 197.

45. Western Union Telegraph Company, *The Story of Western Union* (New York: 1949?), 11, folder 2, box 3, 1993 addendum, collection #205, NMAH. Menahem Blondheim, *News over the Wires: The Telegraph and the Flow of Public Information in America, 1844–1897* (Cambridge, Mass.: Harvard University Press, 1994), 210n5. "Influence of the Telegraph upon Literature," *United States Magazine and Democratic Review* 22 (May 1848), 412.

46. James W. Carey, *Communication as Culture: Essays on Media and Society* (Boston: Unwin Hyman, 1989), 211; Michael Schudson, *The Power of News* (Cambridge, Mass.: Harvard University Press, 1995), 67–68; Robert W. Desmond, *The Information Process: World News Reporting to the Twentieth Century* (Iowa City: University of Iowa Press, 1978), 246–248. William L. Shirer, *Twentieth Century Journey* (New York: Simon and Schuster, 1976), 283–284.

47. Peter D. Eicher, ed., *"Emperor Dead" and Other Historic American Diplomatic Dispatches* (Washington: Congressional Quarterly, 1997), 18–19, 237. On redundant words, see Morgenthau to Bryan, 1 Dec. 1915, 119.2/169; Osborne to Egan, 7 June 1916, 7 June 1916, 119.2/204c; and Osborne to Egan, 28 July 1916, 119.2/211a, all at 1910–29, RG 59, NARA.

48. #57 to Teheran, 30 June 1915, R138730, AA-PA. Mary Adeline Anderson, "Edmund Hammond, Permanent Under Secretary of State for Foreign Affairs, 1854–1873" (Ph. D. diss., London University, 1956), 222. Freiherr von Musulin, *Das*

Haus am Ballplatz: Erinnerungen eines österreich-ungarischen Diplomaten (Munich: Verlag für Kulturpolitik, 1924), 141.

49. For a famous anecdote along these lines, see Lytton Strachey, *Eminent Victorians* (London: Penguin, 1986), 136.

50. Testimony of Andrew Buchanan, 6 May 1861, in *British Parliamentary Papers, Report from the Select Committee on Diplomatic Service, Reports from Committees: 1861,* vol. 6 (ordered by The House of Commons to be printed, 23 July 1861), 130; see also ibid., 147, 308. Extract of *Kölnische Zeitung,* 3 Feb. 1902, R138277, AA-PA.

51. Walewski to French foreign ministry, 17 Nov. 1851, Affaires diverses politiques, 1815–1896, Angleterre, 12, MAE, Paris. Affaires diverses politiques, 1815–1896, Angleterre, 9, MAE, Paris; Cowley to Clarendon, #604, 17 Aug. 1853, FO 519/2, PRO. Walewski to French foreign minister, 19 Mar. and 7 Apr. 1853, Affairs diverses politiques, 1815–1896, Angleterre, 18, MAE, Paris. FO 366/99, PRO.

52. Circulaire, #118, 19 Dec. 1857, Comptabilité, Cartons, 10; Daru, Circulaire, #41, 31 Mar. 1870, Comptabilité, Cartons, 16; and Ribot, Circulaire, #115, 13 June 1890, Comptabilité, Cartons, 34, MAE, Paris. Baeyens, *Au Bout du Quai,* 219.

53. Circular to Her Majesty's ministers abroad, 25 Sept. 1858, in "Papers Relating to the Emoluments of Her Majesty's Foreign Service Messengers Attached to the Foreign Office," *British Parliamentary Papers,* 1859 (sess. 1), 14: 117–118. The French government also expressed concern over the cost of couriers. London embassy to Gramont, 20 June 1870, Comptabilité, 1042, MAE, Nantes.

54. Hammond to the secretary of the treasury, 4 Mar. 1867, T1/6711A, PRO. Vary T. Coates and Bernard Finn et al., *A Retrospective Technology Assessment: Submarine Telegraph—The Transatlantic Cable of 1866* (San Francisco: San Francisco Press, 1979), 87, 89; John J. McCusker, "Comparing the Purchasing Power of Money in the United states (or Colonies) from 1665 to Any Other Year Including the Present," Economic History Services, 2001, *www.eh.net/hmit/ppowerusd/.*

55. Memorandum of Agreement, 15 Jan. 1870, ser. 1, box 1, Collection 73, Anglo-American Telegraph Company Papers, NMAH; Thornton to Clarendon, #24, 17 Jan. 1870, T1/6959A, PRO. Washburne to Hamilton Fish, #1227, 14 Oct. 1875, M34, reel 81, NARA.

56. Hammond to the secretary of the treasury, 4 Mar. 1867, T1/6711A, PRO. Weber, *Masked Dispatches,* 130, 206n5; extract of dispatch from Bruce to Stanley, 8 Jan. 1867, T1/6711A, PRO.

57. Warren Frederick Ilchman, *Professional Diplomacy in the United States, 1779–1939: A Study in Administrative History* (Chicago: University of Chicago Press, 1961), 27–28.

58. W. L. Marcy circular, 10 July 1854, vol. 1, 131; Edwin F. Uhl, 12 Sept. 1894, vol. 4; and W. H. Seward, circular no. 33, 2 Apr. 1863, vol. 1, all at E726, RG 59, NARA.

59. David Paull Nickles, "Telegraph Diplomats: The United States' Relations with France in 1848 and 1870," *Technology and Culture* 40, no. 1 (Jan. 1999), 14–18.

60. *Instructions to the Diplomatic Officers of the United States* (Washington: Department of State, 1896), 100–101, E729; and confidential circulars, 15 Apr. 1898 and 1 June 1898, vol. 5, E726, RG 59, NARA.

61. W. J. Bryan, "Order by the Secretary of State," #58, 4 June 1914; and Wilbur J. Carr, "Telegrams," #323, 29 June 1914, 119.2/78/54, RG 59, 1910–29, NARA. Rachel West, *The Department of State on the Eve of the First World War* (Athens: University of Georgia Press, 1978), 1, 132–133; Mallet, "Mobilization in Hungary," Confidential, 13 July 1914, 763.72/6, M367, reel 12, NARA.

62. General Instruction, 15 Apr. 1920, 119.2/-, 1910–29, RG 59, NARA. Coates and Finn, *Retrospective Technology Assessment,* 87, 89.

63. Raymond A. Jones, *The British Diplomatic Service: 1815–1914* (Gerrards Cross, U.K.: Colin Smythe, 1983), 125. M. S. Anderson, *The Rise of Modern Diplomacy, 1450–1919* (New York: Longman, 1993), 118.

64. Between 1906 and 1908 the Auswärtiges Amt's expenditures for postage and telegrams rose from 694,000 to 894,000 Marks. Magdeburg to imperial chancellor, 12 June 1908, R138280, AA-PA; Schoen circular, 24 Dec. 1908, R139521, AA-PA. After the Spanish-American War the telegraph expenses of the German legation in Washington rose from $3,927.93 in 1898 to $10,888.60 in 1899. Holleben to Hohenlohe-Schillingsfürst, 11 Jan. 1900, R138869, AA-PA.

65. Extract from *Kölnische Zeitung,* 3 Feb. 1903, R138277, AA-PA; B. R. Mitchell, *British Historical Statistics* (Cambridge: Cambridge University Press, 1998); McCusker, "Comparing the Purchasing Power of Money." Zimmermann to the German embassy in Constantinople, 3 Feb. 1915, R138730, AA-PA.

66. Coates and Finn, *Retrospective Technology Assessment,* 87, 89. The price of Atlantic cablegrams declined 98 percent between 1866 and 1904, while the consumer price index fell 44 percent. Samuel H. Williamson, "What Is the Relative Value?" Economic History Services, Apr. 2002, *www.eh.net/hmit/compare/.* Jorma Ahvenainen, "The International Telegraph Union, the Cable Companies, and the Governments," manuscript.

67. Walewski to the French foreign minister, 7 Apr. 1853, ADP, Angleterre, 18, MAE, Paris; Hammond to secretary of the Treasury, 23 July 1867, T1/6711A, and 29 Jan. 1868, T1/6771B, PRO. A. Montgelas to Bülow, 16 June 1909, R138280, AA-PA.

68. Knox, order by the secretary of state, #46, 13 Feb. 1913, 119.2/43, 1910–29, RG 59, NARA. John Hay, circular, 9 June 1904, vol. 6, and Thomas W. Cridler, 21 Dec. 1898, vol. 5, E726, RG 59, NARA; Wilbur J. Carr to American consular officials, #323, 29 June 1914, 119.2/78/54, 1910–29, RG 59, NARA. Huntington Wilson, 1 Mar. 1912 and 14 May 1912, vol. 7, E726, RG 59, NARA.

69. The British Foreign Office reduced its telegraph expenses from £8283 23s in 1859–60 to £2107 in 1867–68. Although these figures show a real trend, their implications are muddled by the fact that Britain adopted a less activist foreign policy during these years. See testimony by Edmund Hammond, 21 Mar. 1870, and "Statement of the Extraordinary Disbursements of Her Majesty's Embassies and Missions Abroad for the Years 1859–60 to 1867–68," in *British Parliamentary Papers, Report*

from the Select Committee on Diplomatic and Consular Services (ordered by The House of Commons to be printed, 25 July 1870), 7: 56, 435. On State Department spending on cablegrams, see RG 59 files, NARA: E272, General and Special Ledgers, 1920–74; and E261, Register and Journals of Contingent Expenses, 1831–53 and 1867–89.

70. Weber, *Masked Dispatches,* 128; deposition of W. H. Seward, 7 July 1870, RG 123 (Records of the U.S. Court of Claims), General Jurisdiction, case file 6151, box 306, NARA.

71. *Petition in the United States Court of Claims: The New York, Newfoundland and London Telegraph Co., vs. The United States,* 10, 25, 40, M179, reel #319, NARA. Depositions of W. H. Seward and Cyrus W. Field, 7 July and 23 Aug. 1870, RG 123, case file 6151, box 306, NARA; *Cases decided in the Court of Claims of the United States at the Dec. Term for 1870,* ed. Charles C. Nott and Samuel H. Huntington, vol. 6 (Washington: Morrison, 1871?), 453; Weber, *Masked Dispatches,* 138, 141, 152–153.

72. Deposition of W. H. Seward, 7 July 1870. Seward to Wilson G. Hunt, 23 Nov. 1867, M40, reel 63, NARA; Weber, *Masked Dispatches,* 148; Seward to Dix, "Circular," 19 Aug. 1867, M77, reel 58, NARA.

73. *Cipher and Key* (1867 code), T1171 (Codebooks of the Department of State, 1867–1876), NARA; Weber, *Masked Dispatches,* 155. E. B. Washburne to Hamilton Fish, #123, 17 Dec. 1869, M34, reel 69, NARA; Fish to Washburne, #95, 1 Jan. 1870, M77, reel 58, NARA. Weber, *U.S. Diplomatic Codes and Ciphers,* 219–220.

74. John Haswell to Hamilton Fish, 8 July 1873, Hamilton Fish Papers, container 95, LC; Weber, *Masked Dispatches,* 157. Weber, *U.S. Diplomatic Codes and Ciphers,* 229, 232, 234–237. Charles Francis Adams, the U.S. minister in London, commented after receiving three messages in the new code, "There has been no slight difficulty experienced in arriving at the meaning of these telegrams." Adams to Seward, #1449, 14 Sept. 1867, M30, reel 90, NARA. Fish to Andrew G. Curtin, 18 Sept. 1871, Fish Papers, box 346, LC.

75. Reverdy Johnson to W. H. Seward, #74, 9 Dec. 1868, M30, reel 93, NARA. Robert Schenck to Hamilton Fish, 9 May 1872, M30, reel 113, NARA; Weber, *U.S. Diplomatic Codes and Ciphers,* 226.

76. W. H. Seward to John C. Deane, 16 Jan. 1868, M40, reel 63, NARA. Daniel Headrick, *The Invisible Weapon: Telecommunications and International Politics, 1851–1945* (Oxford: Oxford University Press, 1991), 225, 227.

77. German diplomats reported that the problem of garbled telegrams forced them to abandon a code designed to reduce length and cut costs. Testimony of Märkl, 26 Mar. 1917, R16919, AA-PA. It also led them to discontinue a system of superencipherment that provided extra security but mutilated messages. Testimony of Heinrich Schafhausen, 9 Aug. 1917, R16919, AA-PA.

78. Harold Temperley, "The Alleged Violations of the Straits Convention by Stratford de Redcliffe between June and September, 1853," *English Historical Review* 49 (1934), 658. "The News by the Caledonia," Washington *Daily Union,* 30 Mar. 1848. "Editorial Etchings," *Cosmopolitan Art Journal* 2, no. 4 (Sept. 1858), 213–214. "The Situation," *New York Herald,* 28 Nov. 1861, 4. Lester G. Lindley, *The Impact of*

the Telegraph on Contract Law (New York: Garland, 1990), ch. 1; James Willard Hurst, *Law and the Conditions of Freedom in the Nineteenth-Century United States* (Madison: University of Wisconsin Press, 1956), 23.

79. Headrick, *Invisible Weapon,* 199–201. For many examples of garbled telegrams, see RM 5/765–766, BA-MA.

80. Lindley, *Impact of the Telegraph,* 30–31; "La Température," *Le Figaro,* 12 Aug. 1895, 1; "Le personnel des télégraphes," *Le Temps,* 5 Aug. 1911; "La Température," *Le Temps,* 22 Aug. 1911, 4; Log Book of Heart's Content Station, folder 1, box 1, ser. 5, collection #73 (Anglo-American Telegraph Company), NMAH. Frank Thomas, *Telefonieren in Deutschland: Organisatoriche, technische und räumliche Entwicklung eines grosstechnischen Systems* (Frankfurt: Campus Verlag, 1995), 161, 166. "Radio vs. Cable vs. Land Wire," #165, box 589, series I, collection 55 (Clark), NMAH. Gramont to Benedetti, telegram of 10 July and letter of 10 July 1870; Benedetti to Gramont, telegram, 10 July 1870, Correspondence Politique, Prusse, 379, MAE, Paris.

81. Bussche to Bethmann Hollweg, 18 Sept. 1915, R2589, AA-PA. Lindley, *Impact of the Telegraph,* 31–32. Kahn, *Codebreakers,* 839.

82. Der Staatssekretär des Reichsmarineamts, Im Auftrage, 5 Nov. 1913, RM5/766, BA-MA. Franklin Mott Gunther to Buck, 10 Mar. 1911, 119.252/137, 1910–29, RG 59, NARA. C. F. Adams to Seward, #1449, 14 Sept. 1867, M30, reel 90, NARA. Alvey A. Adee, 14 Sept. 1900, vol. 5, E726, RG 59, NARA. Karl-Alexander Hampe, *Das Auswärtige Amt in der Ära Bismarck* (Bonn: Bouvier Verlag, 1995), 42–43. On handwritten vs. typewritten telegrams, see the exchange between Arthur Hugh Frazier and William Phillips, 119.2/168, 1910–29, RG 59, NARA; and Weber, *Masked Dispatches,* 211–212.

83. Von Muhlberg to von Treutler, 29 Nov. 1906, Paris Botschaft #1449, AA-PA. *The Cipher of the Department of State* (Washington: GPO, 1876), 5, T1171, NARA. *Cipher and Key,* iii, T1171, NARA.

84. Walewski to Minister of Foreign Affairs, 20 Oct. 1852, ADP, Angleterre, 18, MAE, Paris. Hammond to the secretary of the treasury, 23 July 1867, T1/6711A, PRO. Moran to Fish, #226, 11 Feb. 1871, M30, reel 105, NARA.

85. Baeyens, *Au Bout du Quai,* 217. Matthieu to all code-equipped missions and consulates, 26 Sept. 1911, R138280, AA-PA. Sir Hercules Robinson, charged with implementing Britain's annexation of Fiji in 1874, complained that telegrams from the Colonial Office were frequently so garbled that their "precise purport can often only be conjectured. I suggest that as clearness is of more importance than secrecy in this Fiji matter, it would be as well to telegraph *en clair.*" Robert J. Cain, "Telegraph Cables in the British Empire, 1850–1900" (Ph.D. diss., Duke University, 1970), 196n3. During the July Crisis before the First World War, the Austrian foreign ministry instructed its minister in Belgrade to telegraph news of the Serbian reply in clear text. Albertini, *Origins of the War of 1914,* 2: 374.

86. Weber, *Masked Dispatches,* 206n5. H. F. Taff to Huntington Wilson, 23 Feb. 1912, 119.2/16, 1910–29, RG 59, NARA. T1/7466A, PRO.

87. *Cipher of the Department of State* (1876), 5–6; Weber, *U.S. Diplomatic Codes and Ciphers,* 241. Gramont to Benedetti, 10 July 1870, Correspondence Politique, Prusse, 379, MAE, Paris; J. R. Buck, 22 July 1914, 119.2, RG 59, 1910–29, NARA. Elihu Root, 19 June 1908, vol. 7, E726, RG 59, NARA. Lord Strang et al., *The Foreign Office* (1955; Westport, Conn.: Greenwood, 1984), 151.

88. Bülow, 31 Jan. 1903, R138277, AA-PA. Alvey A. Adee, "Report: Punctuation of Cipher Telegrams," 2 Nov. 1878, E745, RG 59, NARA.

89. Weber, *Masked Dispatches,* 211. Elihu Root, 19 June 1908, vol. 7, E726, RG 59, NARA. For an example of code clerk errors, see telegram to the Admiral Staff of the Navy, 4 Nov. 1912, RM 5/766, BA-MA. Telegrams sent directly tended to contain fewer errors than those sent by more circuitous routes. Curtis Guild, 13 Feb. 1913, 119.2/45, 1910–29, RG 59, NARA.

90. *Verhandlungen des Reichstags: Stenographische Berichte* (vol. 79 in the Harvard University Collection), 4 Dec. 1884, 199–200. Georges Bonnin, ed., *Bismarck and the Hohenzollern Candidature for the Spanish Throne: The Documents in the German Diplomatic Archives* (London: Chatto and Windus, 1957), 195–196, 233–234. On how some technological systems make "operator error" virtually inevitable, see Charles Perrow, *Normal Accidents: Living with High-risk Technologies* (New York: Basic Books, 1984), 9; Scott D. Sagan, *The Limits of Safety: Organizations, Accidents, and Nuclear Weapons* (Princeton, Princeton University Press, 1993), 208.

91. Gauldrée Boilleau to French minister of foreign affairs, "Confidentielle," New York, 14 May 1867, ADP, États-Unis, carton 13, no. 76, MAE, Paris. Dispatch to Bülow, #67, 29 Apr. 1907, R138279, AA-PA. James R. Roosevelt to Thomas Bayard, 31 Nov. 1895, in Charles Callan Tansill, *The Foreign Policy of Thomas F. Bayard, 1885–1897* (New York: Fordham University Press, 1940), 709–715.

92. Gramont to Benedetti, 3 July 1870, Correspondence Politique, Prusse, 379, MAE, Paris. Le Comte Vincent Benedetti, *Ma mission en Prusse* (Paris: Henri Plon, 1871), 327, 336. Alvarez, "Faded Lustre," 124. Lyons to Robert Bunch, "copy," 26 Nov. 1861, FO 5/775, PRO.

93. 11 July 1870, Bonnin, *Bismarck and the Hohenzollern Candidature,* 242. Vittorio Emanuele Orlando, *Memorie (1915–1919)* (Milan: Rizzoli, 1960), 462–463.

94. Some of the allure rubbed off on telegraphers. Nineteenth-century Americans, pointing to such examples as Thomas Edison, Andrew Carnegie, and Theodore Vail (of AT&T), asserted that a knowledge of telegraphy provided upward mobility—this despite the growing proletarianization of the trade. See Edwin Gabler, *The American Telegrapher: A Social History, 1860–1900* (New Brunswick: Rutgers University Press, 1988), 61–63. Women with access to such powerful technology met with curiosity and sometimes suspicion. Jacquelyn Dowd Hall, "O. Delight Smith's Progressive Era," in *Visible Women: New Essays on American Activism,* ed. Nancy A. Hewitt and Suzanne Lebsock (Urbana: University of Illinois Press, 1993), 170–171. But cultural productions, such as stories about men in love with telegraph operators, suggest that female telegraphers also earned a degree of prestige. See Alphonse Allais, "Postes et télégraphes," in *Alphonse Allais: Oeuvres Anthumes* (Paris:

Robert Laffont, 1989), 167–169; Anthony Trollope, "The Telegraph Girl," in *Frau Frohmann and Other Stories* (Leipzig: Bernhard Tauchnitz, 1883).

95. Robert I. Rotberg, *The Founder: Cecil Rhodes and the Pursuit of Power* (New York: Oxford University Press, 1988), 516.

96. Coates and Finn, *Retrospective Technology Assessment,* 92; Baron J. de Szilassy, *Traité pratique de diplomatie moderne* (Paris: Payot, 1928), 165–166. Nicholas Henderson, *The Private Office: A Personal View of Five Foreign Secretaries and of Government from the Inside* (London: Weidenfeld and Nicolson, 1984), 69–70.

CONCLUSION

1. Ernest R. May, *"Lessons" of the Past: The Use and Misuse of History in American Foreign Policy* (New York: Oxford University Press, 1973), 179.

2. For a defense of the practice of responding to unintended effects as they arise rather than attempting to anticipate them, see Herbert A. Simon, "Designing Organizations for an Information-Rich World," in *Computers, Communications, and the Public Interest,* ed. Martin Greenberger (Baltimore: Johns Hopkins University Press, 1971), 47.

3. These definitions are partly based upon the work of Thomas P. Hughes, "Technological Momentum," in *Does Technology Drive History? The Dilemma of Technological Determinism,* ed. Merritt Roe Smith and Leo Marx (Cambridge, Mass.: MIT Press, 1994), 102.

4. Claude Fischer uses the word "structure" for the constraining effect produced by the interaction of many individual decisions. Fischer, *America Calling: A Social History of the Telephone to 1940* (Berkeley: University of California Press, 1992), 19.

5. S. Frederick Starr, "New Communications Technologies and Civil Society," in *Science and the Soviet Social Order,* ed. Loren R. Graham (Cambridge, Mass.: Harvard University Press, 1990), 365n9, 21–22. Ithiel de Sola Pool, *Technologies of Freedom* (Cambridge, Mass.: Harvard University Press, 1983), 5.

6. It appears that social demand drives technological innovation more often than the other way around, but more research on this subject is needed. See Stephen Peter Rosen, *Winning the Next War: Innovation and the Modern Military* (Ithaca: Cornell University Press, 1991), 41–42.

Acknowledgments

I wish to express particular gratitude to Akira Iriye, Ernest R. May, and Stephen Van Evera, whose guidance and advice made this book possible. I am also indebted to David Blackbourn, Rebecca Feldman, Michael Gordon, Nigel Gould Davies, Talbot Imlay, Tom Jehn, Chimène Keitner, Thomas K. McCraw, Susan Pedersen, Bradford Perkins, Melvin Shefftz, James Van Hook, and Bruce Zellers, all of whom provided helpful feedback on sections of the manuscript. Valuable advice came from Alexis Albion, Ted Bromund, Al Brophy, Rebecca Carpenter, Marilina Cirillo, John Collins, Greg Fiete, John Lewis Gaddis, Claudie Gardet, Loren Graham, Robert Gross, Daniel Headrick, Arthur Hoch, Susan Hunt, Haruo Iguchi, Robert Jervis, Paul Kennedy, Mark Kramer, Dan Lindley, Eric Lohr, Erin Mahan, Seth Maislin, Alexis McCrossen, Rose McDermott, Roger Owen, Pat Pearson, Stephen P. Rosen, Daniela Rossini, Jon Shefftz, Zachary Shore, Anthony Silva, Merrit Roe Smith, John Staudenmaier, James Steakley, Zara Steiner, Marisa Szabo, Jack Trumpbour, Jonathan Winkler, and John Womack. This project had its origins in my undergraduate work, which benefited from the encouragement of N. Gordon Levin, John Halsted, James Der Derian, and Hugh Aitken. Camille Smith of Harvard University Press greatly improved the text. Edmund Spevack was a source of inspiration.

It has been a pleasure to work with the staffs of the National Archives, the Library of Congress, the Smithsonian Institution, the Public Record Office in Kew Gardens, the Bundesarchiv-Militärarchiv in Freiburg, the Bundesarchiv in Koblenz, the French foreign ministry archives in Nantes and Paris, the Political Archives of the German foreign ministry (then in Bonn), the Princeton University archives, the Yale University archives, the Harvard University libraries, and the State Department's Bunche Library.

A number of institutions provided support for my project: the John M. Olin

Institute for Strategic Studies, the Center for European Studies, the Charles Warren Center for Studies in American History, the Center for International Affairs, and the History Department, all at Harvard University. I also wish to thank the program in International Security Studies at Yale University, where I received a John M. Olin postdoctoral fellowship.

None of the individuals or institutions I mention are culpable for flaws in this book. Although my current job provides a stimulating environment for thinking about international history, the views expressed here are my own, and do not necessarily represent those of the Department of State.

I dedicate this book to Helen, William, and Laura Nickles.

Index